Advanced Welding Techniques

Mukti Chaturvedi · S. Arungalai Vendan

Advanced Welding Techniques

Holistic View with Design Perspectives

Springer

Mukti Chaturvedi
School of Engineering
Dayananda Sagar University
Bengaluru, Karnataka, India

S. Arungalai Vendan
School of Engineering
Dayananda Sagar University
Bengaluru, Karnataka, India

ISBN 978-981-33-6620-6 ISBN 978-981-33-6621-3 (eBook)
https://doi.org/10.1007/978-981-33-6621-3

This Springer imprint is published by the registered company Springer Nature Singapore Pte Ltd.
The registered company address is: 152 Beach Road, #21-01/04 Gateway East, Singapore 189721, Singapore

Preface

Welding technology has traversed through intense technological evolutions and transformations during the last four decades. The degree of revolution is gargantuan as its application is spreading to multiple co-ordinates of product developments. At present, it appears that welding technology is polarized into two streams: mechanical and metallurgy. The focus on design and power sources is in evanesce as observed from the perception of the welders, trainers and practitioners. Design defines the characteristics that a solution needs to meet the desired success. Design imparts the logical structure to the engineering products for its reliability and performance.

Our interactions with scholars, welding trainers, students and academicians participating in various programs from time to time during the process of preparation of this book has made us believe that there is an inadequacy with the knowledge of design although it is an integral component of welding. The expertise in the field of design for welding is sporadic while many engineers attending to technical glitches on welding of structures to offer solutions are facing conundrum. It is in this context, a need was felt to provide all the basic information on design for welding to the interested groups for their day to today teaching, learning, practicing and subsequent studies. Accordingly, a series of reference materials was initiated by espousing the simple to complex principle: starting with terminology introduction. The sequence of chapters is organized in such a way that the first chapter is conceptually interlinked to understand the second more effectively and so on.

The framework is reflected in the following sequence of production: Introduction to welding processes, design requirements, prominence of design, case studies presenting structural defacements due to inappropriate design, comprehensive surveys on welding processes selected from various process categories, design calculations to be adopted for specific applications and sample calculations. The following objectives were deliberated while initiating this volume: (i) disseminate information and promote in depth knowledge percolation on design vitalities for welding, (ii) awareness creation and motivation to learners to focus on unpolarized streams governing welding which are marginalized in recent times, (iii) to facilitate idea convergence on wide range of experiences/circumstances/problems and topics with adequate insights into up-to-date knowledge.

Though encyclopaedic coverage of this type of work is non-viable in every facet, the provision for scope widening is guaranteed such that information's reported in literatures till 2018 are accommodated for the benefit of learners irrespective of their discipline. A portion of descriptions encompassed, here, are mostly abstracted from the originals. Nevertheless, in few cases, they are acquired only from secondary sources as original were not immediately available. Complex experimental procedures and solutions are presented with minor alterations to complement the objectives or retained as it is to convey with the same technical flavour. Each entry made in this volume is scrutinized to ascertain the intent of benefiting the users to get an overall idea of the descriptions in a quick glance. This book will serve as a linchpin to all those involved in welding science by imparting ample insights to explore this magnificent terrain.

Bengaluru, India

Mukti Chaturvedi
S. Arungalai Vendan

Contents

1 Welding: An Overview 1
1.1 Welding Techniques—Classification 1
1.2 Resistance Welding 2
1.2.1 Process Variations 2
1.2.2 Preferred Materials for RSW 4
1.2.3 Design Considerations 5
1.2.4 Typical Applications of Resistance Weld Variants 5
1.3 Solid-State Welding 5
1.3.1 Process Variations 6
1.3.2 Preferred Materials for Solid-State Welding 7
1.3.3 Design Considerations 8
1.3.4 Typical Applications 8
1.4 Arc Welding 8
1.4.1 Process Variations 9
1.4.2 Preferred Materials for Arc Welding 10
1.4.3 Design Considerations 10
1.4.4 Typical Applications 11
1.5 High-Energy Density Welding 11
1.5.1 Process Variations 11
1.5.2 Preferred Materials for Laser Beam Welding 13
1.5.3 Design Considerations 13
1.5.4 Typical Applications 14
1.6 Conspectus of Welding Techniques 14
References 15

2 Decision Making in Welding Design 17
2.1 Introduction 17
2.2 Standards and Specifications 18
2.3 Factors Affecting Design 18
2.4 Expected Weld Characteristics 19
2.4.1 Example: Design Requirements—For Spot Welding 20

2.5 Effects of Design on Welding Procedure Selection 22
2.6 Weld Distortion and Defects 22
2.6.1 Weld Distortion 22
2.6.2 Weld Defects 23
2.7 Design Methodologies 25
2.7.1 Energy-Based Model 25
2.7.2 Energy Density and Line Energy Estimation 25
2.7.3 Full Factorial Design and Response Surface Methodology Used as a Design of Experiments Approach 26
2.8 Design Failures—Case Study 26
2.8.1 Case History 1: Magnetic Pulse Welding Failure 27
2.8.2 Case History 2: Friction Stir Welding Failure 27
2.8.3 Case History 3: Laser Welding Failure 27
2.8.4 Case History 4: TIG Welding Failure 28
2.8.5 Case History 5: Bridge Failure 28
2.8.6 Case History 6: Chernobyl Reactor Accident 29
2.8.7 Case History 8: Aircraft Engine Failure 30
2.9 Conspectus of Weld Design Studies 30
References 33

3 Resistance Spot Welding and Design 35
3.1 Introduction 35
3.2 Process Applications 39
3.3 Compatible Materials 39
3.4 Fundamentals of Resistance Spot Weld Process 39
3.5 Resistance Spot Welding Machine Details 42
3.6 Literature Review 46
3.7 Sample Design Data, Process Parameters and Design Calculations 55
3.7.1 Case Study for Design Calculations 55
3.7.2 Sample Data for a Particular Application 59
3.8 Conspectus of Design Studies in Resistance Spot Welding 60
References 60

4 Tungsten Inert Gas Welding and Design 63
4.1 Introduction 63
4.2 Process Applications 64
4.3 Compatible Materials 65
4.4 Fundamentals of Tungsten Inert Gas Weld Process 66
4.5 Tungsten Inert Gas Welding Machine Details 68
4.5.1 Power Source 69
4.5.2 Torch 70
4.5.3 Electrodes for Welding 71
4.5.4 Shielding Gases 72
4.6 Literature Survey 73

4.7 Sample Design Data, Process Parameters, and Design Calculations 77
4.7.1 Welding Amperage Selection 79
4.7.2 Sample Design Calculations 80
4.7.3 Sample Data for Various Materials and Applications 85
4.8 Conspectus of Design in Tungsten Inert Gas Welding 87
References 87

5 Laser Beam Welding and Design 89
5.1 Introduction 89
5.2 Process Applications 92
5.3 Compatible Materials 92
5.4 Fundamentals of Laser Beam Weld Process 94
5.5 Laser Welding Machine Details 94
5.6 Literature Survey 97
5.7 Sample Design Data, Process Parameters, and Design Calculations 102
5.7.1 Design Sequence in Welding 103
5.7.2 Approach I: Heat Affected Zone Considerations 104
5.7.3 Approach II: Thermal Gradients with Varying Thickness 106
5.7.4 Approach III: Use of Energy-Based Model for Weld Depth Determination 109
5.7.5 Approach IV: Use of Process Optimization Techniques 113
5.7.6 Sample Parameter Values 115
5.7.7 Case Study for Selection of Laser Source 119
5.7.8 Laser Welding Parameters 122
5.7.9 Power Supply for Laser Machine 125
5.8 Conspectus of Design Studies in Laser Beam Welding 129
References 130

6 Friction Stir Welding and Design 133
6.1 Introduction 133
6.2 Process Applications 137
6.3 Compatible Materials 138
6.4 Fundamentals of Friction Stir Weld Process 138
6.4.1 Heat Generation Analysis 139
6.5 Friction Stir Welding Machine Details 142
6.6 Literature Survey 147
6.7 Sample Design Data, Process Parameters & Design Calculations 153
6.7.1 Sample Data I 154
6.7.2 Design Calculations- Case Study 157
6.8 Conspectus of Design Studies in Friction Stir Welding 164
References 164

7 Magnetic Pulse Welding and Design 167
7.1 Introduction 167
7.2 Process Applications 170
7.3 Compatible Materials 170
7.4 Fundamentals of Magnetic Pulse Weld Process 170
7.5 Magnetic Pulse Welding Machine Details 173
7.6 Literature Survey 175
7.7 Sample Design Data, Process Parameters, and Design Calculations 180
7.7.1 Design Calculations I 181
7.7.2 Design Calculations II 185
7.7.3 Design Case Study III 187
7.7.4 Design Case Study IV 189
7.7.5 Design Calculations—Weldability Curve 191
7.7.6 MPW: Electrical Model 193
7.8 Conspectus of Design Studies in Magnetic Pulse Welding 196
References 196

About the Authors

Mukti Chaturvedi is Assistant Professor in the School of Engineering at Dayananda Sagar University. She graduated with electrical engineering and did postgraduation in VLSI and Embedded Systems. During her graduation, she cultivated interest in power electronics and control system. She instituted various projects pertinent to power electronics, control system and electrical drives. Relentless efforts resulted in technical acquaintance with several concepts from versatile fields encompassing manufacturing, especially welding. With the opportunity to work in various industries, she has got the exposure to different domains of work. The current role as a faculty member provided ample scope for reading, learning, teaching and performing interesting experimentations on different subjects moulding her to acquire multifaceted outlook. This led her to identify welding technology to be her area for Ph.D. which constitutes power source design, configurations and process understanding, automations through programming for control and converter modules.

S. Arungalai Vendan is presently Associate Professor in the School of Engineering at Dayananda Sagar University, Bangalore. Previously, he was a faculty member in Industrial Automation and Instrumentation Division at VIT Vellore. He undertook research on advanced welding processes since 2006. He received his Ph.D. from the National Institute of Technology, Tiruchirappalli, India, in 2010. He has received several fellowships and awards for his technical contributions from various government and private organizations. He has successfully completed numerous government-funded research projects and industrial consultancy tasks and has published more than 80 research papers in reputed international journals and conference proceedings. He has associations with top manufacturing industries and research and development centres under various capacities. His research interest mainly focuses on the interdisciplinary science underlying welding which includes the confluence of terminologies from electrical/mechanical/metallurgical materials and magnetic streams.

List of Figures

Fig. 1.1 Classification of welding processes 2
Fig. 1.2 Schematic and weld schedule of RSW 3
Fig. 1.3 Types of Resistance Welding Methods 3
Fig. 1.4 Process variations of solid-state weld 6
Fig. 1.5 Schematic of arc welding 9
Fig. 1.6 Schematic of non-consumable electrode arc welding 10
Fig. 1.7 Schematic of laser beam welding 12
Fig. 1.8 Components of Nd:YAG Laser 12
Fig. 1.9 **a** Fiber laser mechanism, **b** fiber 13
Fig. 2.1 Plates of different thickness 20
Fig. 2.2 Schematic showing distance of form to the center of spot weld 21
Fig. 2.3 Schematic for overlap region in spot weld 21
Fig. 2.4 Part of lorry having weld defect 28
Fig. 2.5 Weld joint between carbon steel piping and oil tank 29
Fig. 2.6 Arrangement of the core of Chernobyl unit 4 31
Fig. 2.7 Annulurs in an aircraft engine 32
Fig. 3.1 Schematic of RSW circuit 36
Fig. 3.2 Cross section of RSW nugget 37
Fig. 3.3 Effect of voltage on load bearing capacity 38
Fig. 3.4 Spot weld lobe curve 38
Fig. 3.5 Various resistances and theoretical dynamic resistance curve 41
Fig. 3.6 Machine and control link 43
Fig. 3.7 Schematic of a capacitive discharge welder 43
Fig. 3.8 AC RSW supply 44
Fig. 3.9 HFDC power supply circuit 45
Fig. 3.10 Welding current for different power supply mechanism 45
Fig. 3.11 Sites of failure initiation 46
Fig. 3.12 Dependence of tensile–shear force on weld time and welding force 47
Fig. 3.13 Schematic for electrode tip voltage measurement circuit 48

Fig. 3.14 Weld current sensing circuit using hall-effect sensor. 48
Fig. 3.15 Temperature distribution 49
Fig. 3.16 Nugget dimensions for one sample 49
Fig. 3.17 SEPIC converter circuit 50
Fig. 3.18 Synchronous transistor in the buck converter circuit 50
Fig. 3.19 Block diagram for ISCC strategy 51
Fig. 3.20 Inverter-type spot welding circuit 52
Fig. 3.21 Fuse placement for inverter-based power supply. 52
Fig. 3.22 Parametric model of RSW and weld cap geometry parameters. 53
Fig. 3.23 Weld geometry for unequal thickness sheets. 54
Fig. 3.24 Stress distribution along electrode workpiece interface 54
Fig. 3.25 Standard buck converter 56
Fig. 4.1 Schematic of TIG welding 64
Fig. 4.2 Configuration possible with TIG welding 64
Fig. 4.3 TIG weld process 66
Fig. 4.4 Constant current power source 69
Fig. 4.5 TIG torch 71
Fig. 4.6 Control of the AC waveform 72
Fig. 4.7 Slope current control during start and stop 73
Fig. 4.8 Screw and nut mechanism assembly 74
Fig. 4.9 Temperature distribution during TIG process 74
Fig. 4.10 Stress distribution during TIG process 75
Fig. 4.11 Schematic of process showing filler materials placed in a groove 76
Fig. 4.12 Influence of weld speed, current, and gas flow rate on bead height 78
Fig. 4.13 Influence of weld speed, current and gas flow rate on bead penetration. 78
Fig. 4.14 Influence of weld speed, current and gas flow rate on bead width 79
Fig. 4.15 Forces influencing the weld pool 83
Fig. 4.16 Pulsed current time waveform. 84
Fig. 5.1 Laser welding schematic 90
Fig. 5.2 Relationship between power density and weld depth 91
Fig. 5.3 Schematic of conduction limited and keyhole welding 91
Fig. 5.4 Block arrangement for laser machine 95
Fig. 5.5 Gas laser schematic. 95
Fig. 5.6 Arrangement of Nd:YAG laser 96
Fig. 5.7 Power density versus interaction time for various laser processes. 98
Fig. 5.8 Weld dimensions versus laser energy 99

Fig. 5.9 **a** Weld dimensions versus incident angle, **b** weld pool volume versus incident angle ... 99
Fig. 5.10 Joining speeds as function of used laser power for various optical setups and spot diameter ... 100
Fig. 5.11 Experimental and ANN estimated values for different experimental conditions ... 101
Fig. 5.12 Hybrid laser arc welding ... 101
Fig. 5.13 Weld cross section and assumed molten pool volume ... 104
Fig. 5.14 Structural areas of HAZ ... 105
Fig. 5.15 Two-dimensional conduction mode laser beam ... 106
Fig. 5.16 Temperature distribution for thin plate ... 107
Fig. 5.17 Temperature distribution for thin plate ... 108
Fig. 5.18 Cooling rates with variation in plate thickness ... 109
Fig. 5.19 Reinforcement form factor versus energy density ... 112
Fig. 5.20 Variation of penetration size factor with energy density ... 113
Fig. 5.21 Plots showing effect of process parameters on UTS and hardness ... 116
Fig. 5.22 Relation between DOP, pulse off time, and duty cycle ... 118
Fig. 5.23 Laser beam focused on the weld metal ... 121
Fig. 5.24 Cross section of pulsed laser seam weld ... 122
Fig. 5.25 Calibration curve of laser machine ... 124
Fig. 5.26 Pulse shape of the flashlamp current ... 126
Fig. 5.27 Circuit for flashlamp power supply ... 126
Fig. 5.28 Charging circuit ... 127
Fig. 5.29 Overvoltage circuit ... 128
Fig. 5.30 Flashlamp RLC circuit ... 129
Fig. 5.31 Response of critically damped circuit ... 129
Fig. 6.1 Schematic of friction stir weld process ... 134
Fig. 6.2 FSW tool shape ... 135
Fig. 6.3 Tool and workpiece tilted with respect to each other ... 136
Fig. 6.4 Forces acting in FSW and exit hole ... 136
Fig. 6.5 Variation of stress with temperature ... 137
Fig. 6.6 Possible configurations, **a** T-joint, **b** corner joint ... 138
Fig. 6.7 Schematic of FSW tool ... 139
Fig. 6.8 Schematic of surface orientations and infinitesimal segment areas, **a** concave shoulder, **b** pin side, **c** pin tip ... 140
Fig. 6.9 Conventional FSW fixture requirements ... 142
Fig. 6.10 Bobbin FSW tool ... 143
Fig. 6.11 Self-reacting Bobbin tool ... 143
Fig. 6.12 Fixed gap Bobbin tool ... 144
Fig. 6.13 Microstructural classification for FSW product ... 146
Fig. 6.14 Effect of PID gains on temperature response ... 148
Fig. 6.15 Torque, spindle speed power, and temperature response to step changes in desired temperature ... 148

Fig. 6.16 Variation of energy and torque requirement with welding speed and rotational speed. Dashed line—experimental values reported and the solid line—numerically calculated values 149
Fig. 6.17 FOPDT model 150
Fig. 6.18 Tool–workpiece thermocouple method 151
Fig. 6.19 A typical monitoring architecture 151
Fig. 6.20 Offline control mechanism for design and material parameters 152
Fig. 6.21 Transverse view and detail of the parameters 153
Fig. 6.22 Comparison curve of temperature with and without assisted heating 154
Fig. 6.23 Effect of process parameters on various forces 156
Fig. 6.24 Specific weld energy versus weld speed curve 156
Fig. 6.25 Tool geometry 158
Fig. 6.26 Friction power versus contact pressure 159
Fig. 6.27 Plastic deformation energy as a function of welding speed for different shear stress values 160
Fig. 6.28 Effect of welding speed on power Sources generated during FSW 160
Fig. 6.29 Comparison of power versus weld speed at melting temperature and recrystallisation temperature 161
Fig. 6.30 Variation of power with weld speed at Tm for different weld widths 162
Fig. 6.31 Power versus radius ratio 162
Fig. 7.1 **a** Schematic of current discharge circuit, **b** middle section close up 168
Fig. 7.2 MPW arrangement showing electromagnetic effects 169
Fig. 7.3 MPW schematic 169
Fig. 7.4 Discharge current and magnetic pressure variation with time 171
Fig. 7.5 Sample showing center of welding joint 171
Fig. 7.6 Variation of shear strength with standoff distance 172
Fig. 7.7 Skin depth and resistivity relation 173
Fig. 7.8 MPW system 174
Fig. 7.9 MPW setup for tubes 174
Fig. 7.10 Current waveform in an underdamped circuit 175
Fig. 7.11 Effect of coil width on magnetic pressure 176
Fig. 7.12 Experimental results for various axial positions and variation of time 177
Fig. 7.13 Effect of varying influencing parameters on the impacting conditions 178
Fig. 7.14 Variation of skin depth with frequency 178

Fig. 7.15 CAD model for the working coil with tubular workpieces 179
Fig. 7.16 Simulation results for Al-SS workpieces with Cu coil 180
Fig. 7.17 Dependence of field shaper function on length of working zone 182
Fig. 7.18 Influence of applied voltage, plate thickness, and standoff distance on impact velocity. 186
Fig. 7.19 Schematic of MPW discharge circuit 187
Fig. 7.20 3D model of the coil. 188
Fig. 7.21 Current waveform at 9 kJ discharge energy 188
Fig. 7.22 Interface between Al and Fe/Ti/Mg 189
Fig. 7.23 Variation of welded surface-to-contact surface ratio 190
Fig. 7.24 Bonding interface 190
Fig. 7.25 Weldability window for 6061 T0 Al alloy 191
Fig. 7.26 Schematic of analytical process. 193
Fig. 7.27 Electrical model 194

List of Tables

Table 2.1 Preferred weld method for different weld specifications 22
Table 3.1 Material properties of AISI 316L and DSS 2205 56
Table 4.1 Properties and choice of AC/DC power source for TIG welding 70
Table 4.2 Parameter values for manual TIG welding of AL using HF AC supply 80
Table 4.3 Parameter values for manual TIG welding of MS using DC supply 81
Table 4.4 Observation table for input parameters and the resulting bead width 86
Table 5.1 Materials and their properties for the weld process 93
Table 5.2 Sample calculations for the use of optimization techniques 115
Table 5.3 Optimum values of the variable parameters 116
Table 5.4 UTS and Brinell hardness for given parameter values 116
Table 5.5 Process parameters chosen for SET II: laser welding of Austenitic 304L stainless steel sheet 117
Table 5.6 Mechanical properties of DP1000 123
Table 5.7 Properties of laser system: SISMA SWA 300 123
Table 6.1 Design features and effects of the FSW tool 145
Table 6.2 Weld zones and their characteristics. 146
Table 6.3 Mechanical properties of Al2014-T3 157

Chapter 1
Welding: An Overview

Welding involves binding of two materials to form a single component. The materials to be joined are subjected to heat or pressure or combination of both. The material prior to welding needs to be qualified as weldable which indicates that welds of sufficient size and strength can be obtained for the material when welded with standard welding equipment and procedures.

Advent of versatile materials catering to the twenty-first century industrial requirements attributed to the evolution of advanced welding technologies that relies on complex phenomena.

A portion of this book covers the design procedures for selecting optimal ranges of process parameters in the different techniques of welding. Attainment of optimal ranges for these parameters is imperative to yield a weld with good quality and strength.

1.1 Welding Techniques—Classification

Fig. 1.1 shows the classification of the welding techniques based on the type of heat source. Different techniques are thus governed by various process parameters depending upon the material properties and the application of the weld product. Based on the type of heat source, the operating principles need to be well understood for working in any stage of the weld process. The following section describes the operating principles of various welding categories covering the multidisciplinary aspects of each process and the involved design parameters in brief.

Each process has been covered to the complete detail in the following chapters. The techniques have been chosen from the categories of resistance, arc, solid state and high-energy density welding processes. These processes allow for better control and also can be optimized with possible automation.

M. Chaturvedi and S. Arungalai Vendan, *Advanced Welding Techniques*,
https://doi.org/10.1007/978-981-33-6621-3_1

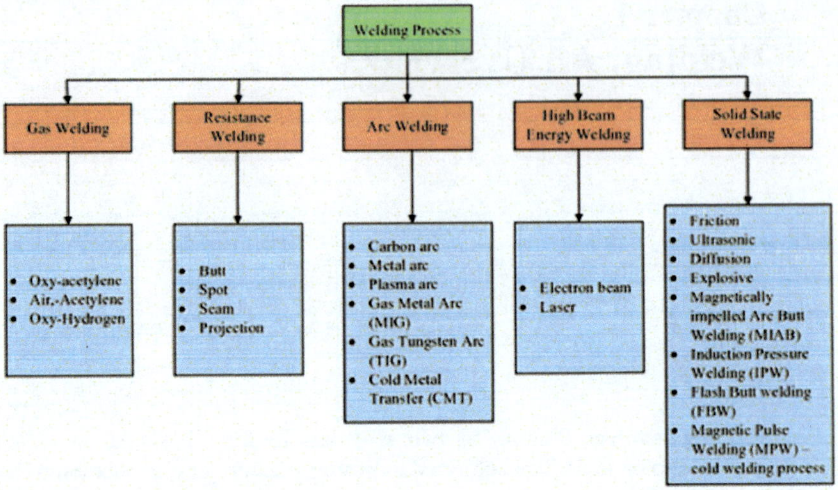

Fig. 1.1 Classification of welding processes

1.2 Resistance Welding

This process involves joining of two metals with the help of resistance offered to the flow of current in a material. The joule heating at the common surface of the weld materials causes the metals to undergo localized heating and consequent melting, and thus, they join together. This heat generation is affected by the resistance at any point in the circuit. Current level, electrode force, and materials being welded are the variables expected to cause significant variation in the shape of the weld. The current flow duration and the electrode pressure need to be controlled for achieving the weld characteristics. Applying optimal force consistently improves the material joining as it reduces the path resistance and also eliminates the oxide layer in the surface [1].

The basic weld schedule and schematic for resistance welding are shown in Fig. 1.2. The weld schedule describes the stages involved in the RSW process described as: holding the elements (squeeze time), heating (weld time), and the hold time. The electrode force acts on the materials throughout the process.

1.2.1 Process Variations

Resistance welding techniques are widely used in automatic pressure welding machines on factory automation sites. Figure 1.3 below describes the types of resistance welding techniques [2].

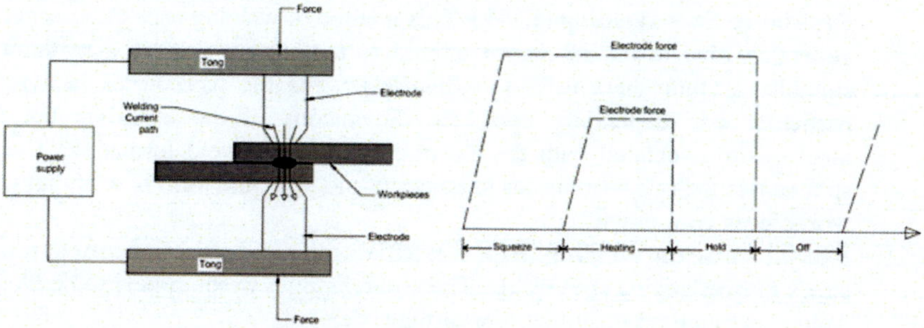

Fig. 1.2 Schematic and weld schedule of RSW

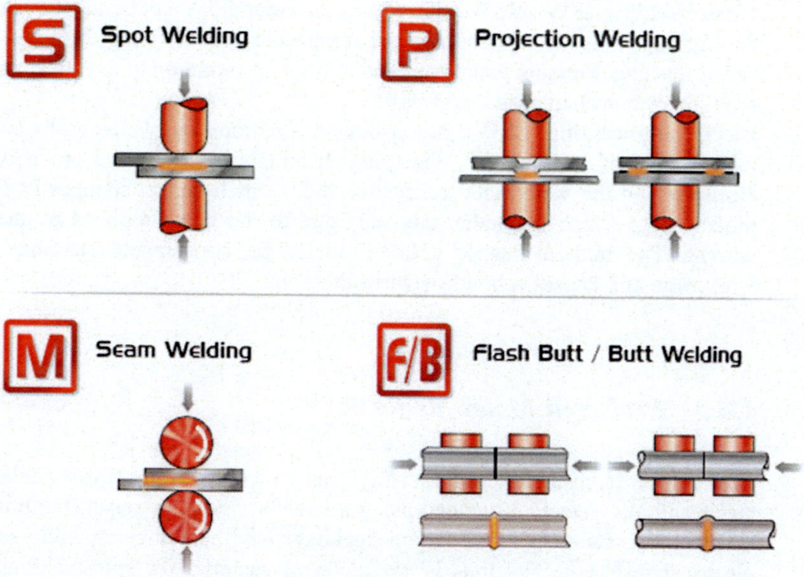

Fig. 1.3 Types of Resistance Welding Methods [2]. *Source* https://www.welding-machine-dahching.com/about-welding.html

Resistance Spot Welding (RSW): This method of welding uses the concept of heat generation due to the resistance offered to the current flow in a material and is suitable for thin sheet welding. This heat melts the workpieces in a spot of a particular size depending upon the dimensions of the electrode tip. Control mechanism associated with the electrode guides the weld formation. The melted spot causes the two workpieces to coalesce and takes the form of a nugget when the workpieces cool down.
Resistance Seam Welding (RSEW): RSW with rapid pulses of current results in a series of overlapping spot welds. These overlapping welds appear to be like a seam weld. This method is suitable for airtight welding.
Resistance Projection Welding (RPW): Localized RSW with welding intended at projections on the component and the sheet metal which are clamped between current carrying plates. Electrodes used in this welding must retain the hardness value during normal temperature.
Flash Welding (FW): RSW with voltage of around 5 V applied at the clamps and the high spots at the contact area are removed by deoxidizing the joint which is called flashing. Pressure is then applied to forge the weld on to the thick workpieces such as anchor chain, rails, and pipes.
Electroslag welding (ESW): A consumable electrode and welding flux are used in this process of welding [3]. The joule heating caused by the arc between the electrode and the weld metal causes the flux to melt, and it changes its form to a molten slag, which maintains this state due to the heat produced by the electric current. The molten slag at 3500 °F melts the consumable electrode and the workpiece and causes a bond between the two.

1.2.2 Preferred Materials for RSW

Low-carbon steels and aluminum alloys can be welded using RSW. Higher-carbon steels and alloy steels, if welded using this technique, may result in a brittle weld.

Steel and specifically low-carbon steel have low thermal conductivity and higher electrical resistance and thus is the preferred material for spot welding. Current requirement in zinc-coated galvanized steel is larger than uncoated steels. Quality of weld surface is degraded if copper electrode is used with zinc alloys.

Nickel and super austenitic alloys can also be used in limited cases because of the tendency to crack due to absence of ferrite in these alloys. Special consideration should be taken in the form of wider joint angles, lower heat input, and flat bead shapes.

1.2.3 Design Considerations

- Typical joint designs that are possible with resistance weld are:
  - Lap resistance spot weld and resistance seam weld
  - Edge RSEW
  - Butt flash weld and electroslag weld.
- Diameter of the spot in spot welding should be significantly higher than the material thickness.
- For the processes of spot and seam welding:
  - Workable sheet thickness is in the range of 3–6 mm.
  - Weld can be done for up to 3:1 thickness ratio of the weld pieces.
- For flash welding,
  - The weld pieces should have same cross section
  - Workpiece may be 0.2-mm-thick sheets or sections up to 0.1 m^2 in area.
- Slag welding can be applied to same thickness sheets in the size range of 25–500 mm
- Vertical welds can restrict design freedom in ESW.

1.2.4 Typical Applications of Resistance Weld Variants

Spot welding: automobiles, aviation, domestic applications, other lightweight structures [2, 3]
Seam welding: automobile components—fuel tanks, radiators, and making of cans
Projection weld: reinforcing rings, captive nuts, pins, and studs to sheet metal, wire mesh
Flash welding: for components having uniform cross sections
Slag weld: On-site fabrication of structural components in buildings and bridges
Coated sheet metals can be welded with this resistance weld processes, except the electroslag weld.

1.3 Solid-State Welding

Solid-state welding—In this method, the material to be welded does not get melted; instead, it is only taken up to its solidus temperature, where it gets deformed at the interface causing formation of bond at the interface. The plastic deformation of the weld pieces which further affects the weld is caused by utilizing pressure, friction heat, or high-energy impact.

1.3.1 Process Variations

Schematic of some of the process variations are shown in Fig. 1.4 [2, 3].

- Cold welding: Plastic deformation of materials is achieved at room temperature with the application of various forces at the weld interface. Cold pressure spot welding is effective for welding of sheet metals.
- Ultrasonic welding: High-speed oscillating vibrations are applied at the weld interface with the electrode. This disrupts the surface oxides for effective cleaning of the weld surface. Induced friction causes heating up to deformation stage and causes the bond in the two materials. Spot welding can also be implemented using this concept. Some of the variants and their applications are:

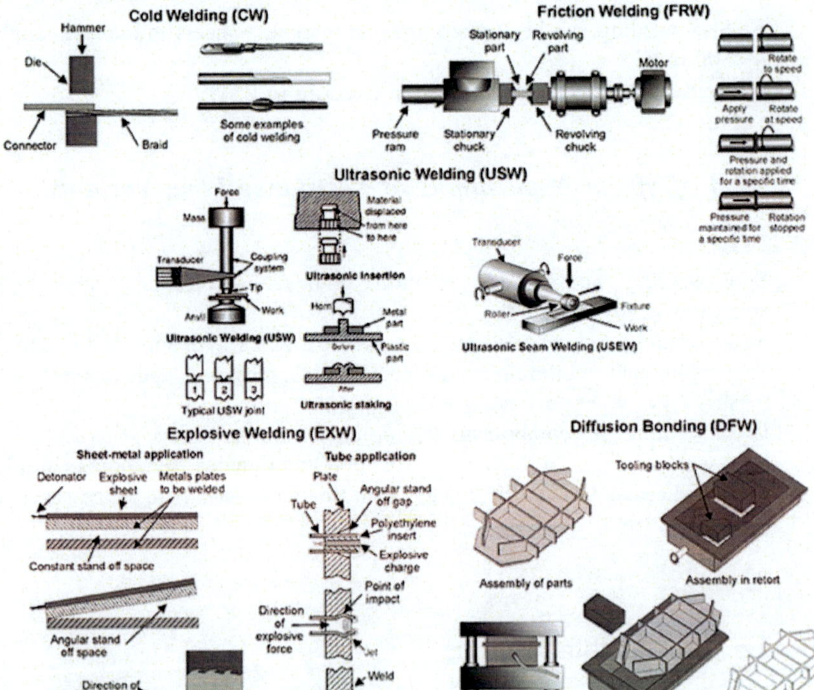

Fig. 1.4 Process variations of solid-state weld [2]

- Seam welding: results in a seam weld with the help of a roller that moves along the weld interface.
- Soldering: Localized heating is caused by high-frequency oscillations of the electrode at the joint surface. No flux material is required for this method, but premachining may be required to have uniform surfaces.
- Insertion: This is used to forge plastic with metal inserts for any required fastening.
- Staking: used for light plastic assemblies.

- Friction welding: Bonding between two weld pieces is obtained with the frictional heat generated between the rotating and stationary weld pieces. Weld is affected by the plastic deformation caused due to frictional heat and pressure applied at the interface.
  - Friction stir welding: A rotating tool is moved along the interface of the two stationary parts, thus creating heat and the transformation to solidus state. Non-consumable material is chosen as the electrode which has rotational and transverse movement along the weld line. Design and material of the tool are critical to the formation of an efficient weld.
- Explosion welding: used for joining sheet metals or tubes by causing an energy transfer with the help of explosive charge which results in high-energy impact on one by the other sheet. This causes plastic deformation and thus resulting in a bond. The interlocking bond has a wavy pattern which results in a strong mechanical bond.
- Diffusion welding: Localized plastic deformation at the weld surface is caused by bringing together the weld pieces in an inert atmosphere and subsequent application of moderate pressure and temperature. Diffusion of atoms between the two surfaces causes the coalescence to occur.

1.3.2 Preferred Materials for Solid-State Welding

- Cold welding: ductile metals such as carbon steels, aluminum, copper, and precious metals.
- Friction welding: can be used for thermoplastics and refractory metal,
- Ultrasonic welding: Al–Cu alloys, carbon steels, limited thermoplastics, and ductile metals.
- Explosive welding: used for welding carbon steels, aluminum, copper, and titanium alloys.
- Diffusion bonding: Cu–Mg alloys, steel stainless and low alloy mixtures, Al, titanium, and precious metals.

1.3.3 Design Considerations

Welding of pieces having unequal thickness can be performed over a range of thicknesses and the type of joint design for each variant as listed below:

- Cold welding: 5–20 mm, lap and butt weld
- Ultrasonic weld: 0.1–3 mm, lap and edge weld
- Explosion weld: 20–500 mm, maximum surface area = 20 m^2, lap, butt, and flange weld
- Friction welding: diameter range: 2–150 mm and maximum surface are of up to 0.02 m^2. The weld pieces should have rotational symmetry. Butt, rabbet, lap, fillet, and hem weld
- Diffusion weld: 0.5–20 mm, lap and T-joint weld.

1.3.4 Typical Applications

- Cold welding: used for joining caps to tubes, electrical terminations, and cable joining.
- Ultrasonic welding: applied for light assembly work of either sheet metal or plastics and electrical equipment.
- Explosive welding: In process industry, petrochemical industry, marine applications for cladding and to improve corrosion resistance.
- Diffusion welding: aerospace, nuclear, and biomedical industries for high-strength structural components and implants and also applied in electrical devices for making metal laminates for electrical devices.
- Friction welding: applied in automotive industry for gear assemblies, and for the joining of hub-ends to axle casings, valve stems to heads.
- Friction stir welding and magnetic pulse welding are used in automotive, aerospace, and electronics industries to accomplish strong and lightweight components and assemblies. Joining of components with vastly different material properties, which is not feasible through fusion welding, may be achieved with magnetic pulse welding.

1.4 Arc Welding

An arc created with an appropriate power supply causes this weld to take effect. It is a fusion process, in which the energy transfer due to the arc as heat causes the melting of the workpieces at the interface and subsequent bonding between the components. Arrangement is shown in Fig. 1.5. DC or AC supply may be used to

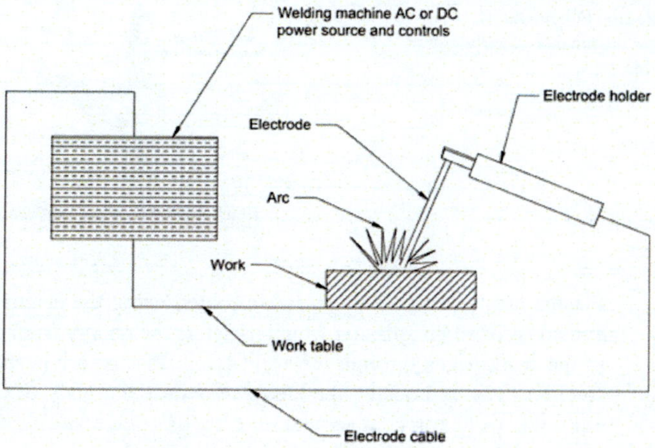

Fig. 1.5 Schematic of arc welding

connect to the electrode in either positive or negative polarities. The electrode used may be consumable or non-consumable electrode. The heat created due to the arc raises the temperature to around 6500 °F at which the joint melts [2].

Few metals with lesser ionization potential, when get heated up by the arc to high temperatures, tend to react with the atmospheric gases. Thus, a protective shielding gas or slag is used to protect the molten workpieces against the atmospheric effects. The interface, in the molten state, on cooling, solidifies into the required weld.

1.4.1 Process Variations

Arc welding processes can be classified based on electrode types in use as consumable or non-consumable.

- **Consumable Electrode Processes**: These are the processes in which the electrode gets consumed during joining. Shielded metal, gas metal, flux cored, and submerged arc welding are few examples of these processes.
  - The parent metal is melted, and the weld is created with the continuous feed of the wire as in Fig. 1.5. Argon or CO_2 is used as a shielding gas.
- **Non-consumable Electrode Processes**: These are the processes in which the electrode does not get consumed during the joining as in Fig. 1.6. Some of these processes are tungsten inert gas (TIG) or gas tungsten arc welding (GTAW), plasma arc welding [4].

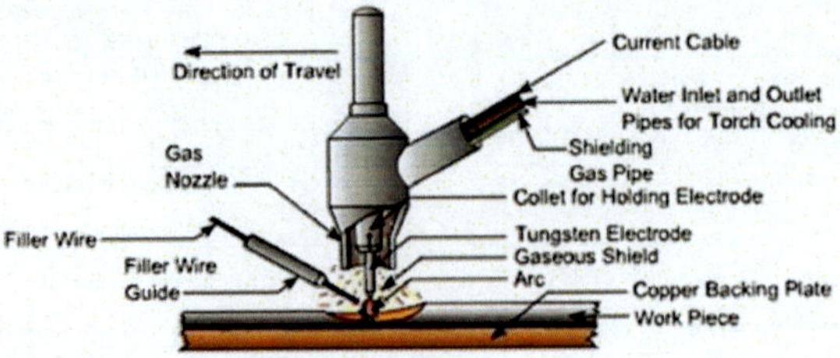

Fig. 1.6 Schematic of non-consumable electrode arc welding [5]

- Plasma arc welding: Two parts are joined using the electrically conductive and constricted gas-plasma which transfers the energy from the power source to the workpieces through the weld torch. Plasma arc is separated from the shielding gas by housing the electrode within the body of the weld torch. It exits the weld torch at temperature in the range of 50,000 °F creating a plasma column which causes formation of melt pool in the workpieces.
- TIG is used in the fabrication of components in fusion reactor. It uses a non-consumable electrode and a separate filler metal with an inert shielding gas. The intense, but small arc provided by the pointed electrode produces good quality and precise welds.

1.4.2 Preferred Materials for Arc Welding

MIG, TIG—These processes are suitable for almost all metals and alloys. Welding of dissimilar materials with TIG and MIG is complex and poses technical challenges. These processes are predominantly used for joining low alloy carbon steel, stainless steel, aluminium, nickel, magnesium, titanium alloys and copper. Refractory alloys and cast iron can also be welded [2, 4, 5].

1.4.3 Design Considerations

- All types of joints—butt, corner, edge, lap, and fillet—are appropriate for MIG and TIG welding.
- MIG—gives an efficient weld for vertical and overhead applications.
- TIG welding is suited to most welding positions, but horizontal welding orientations are preferred during designing.
- MIG with short-circuiting metal transfer is recommended for steels from about 0.250 in. (6.35 mm) thick down to about 0.020 in. (0.51 mm).
- The pulsed arc method is appropriate for sheets of 0.048 in. (1.22 mm). In contrast, TIG can be used to weld sheet as thin as 0.005 in. (0.13 mm).

- Thin sections can be joined to thicker sections more readily than with resistance welding.
  - Minimum sheet thickness = 0.5 mm (6 mm for cast iron).
  - Maximum thickness, generally,

    Carbon, low alloy, and stainless steels; cast iron, aluminum, magnesium, nickel, titanium alloys, and copper = 80 mm
    Refractory alloys = 6 mm.
  - Multiple weld runs required on sheet thicknesses ≥ 5 mm.
- TIG: suitable for sheet thickness in the range of 0.2–6 mm. The maximum thickness limit for copper and refractory alloys is 3 mm, whereas that for low alloy carbon and stainless steels is 6 mm. A decline in the production rate due to the requirement of multiple weld runs is observed in case of heavier gauges of aluminum and titanium of up to 15 mm.

1.4.4 Typical Applications

Arc welding process is a commonly used process in the automotive, aerospace, oil and gas, power industries, and in construction sectors.

MIG applications: general fabrication, structural steel work, and automobile bodywork.

TIG applications: found in nuclear plant components, chemical plant pipe work, structural components in aviation industry, and also for sheet metal fabrication.

1.5 High-Energy Density Welding

Laser welding—Fusion takes place by the absorption of a high-power density narrow beam of light. Focusing of the light generally termed laser is performed by mirrors or lenses Fig. 1.7.

1.5.1 Process Variations

Laser welding may be accomplished using the pulsed neodymium-doped yttrium aluminum garnet (Nd:YAG) or fiber or diode. Selection of laser source is done based upon the application requirements. Pulsed and continuous wave modes can be used, depending on economics of process.

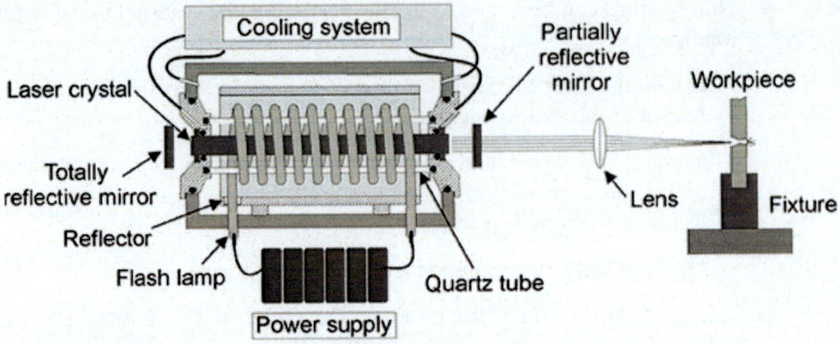

Fig. 1.7 Schematic of laser beam welding [2]

Pulsed Nd:YAG laser is most commonly used technique for microwelding of implantable devices. Major components of Nd:YAG laser are shown in Fig. 1.8.

Fiber laser architectures may be scaled for various weld dimensions and used for high-speed seam welding and are efficient. Mechanism of fiber laser is shown in Fig. 1.9.

Diode laser technique are opted for welding plastics in automotive industry, while high-power level diode lasers are used to weld metals.

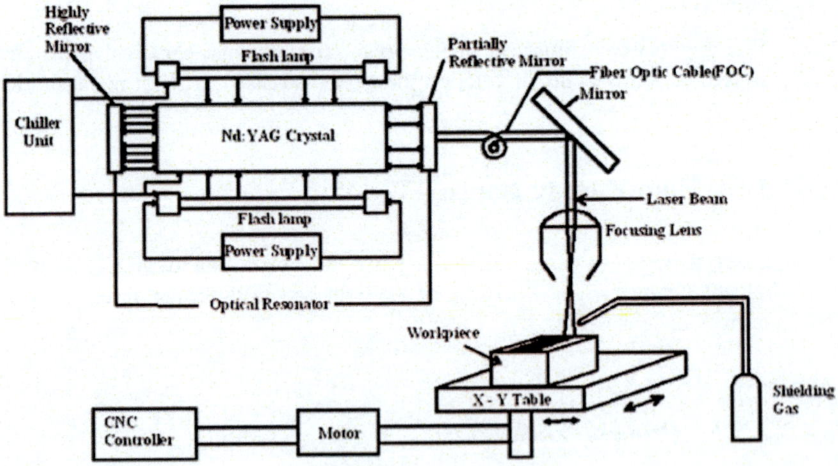

Fig. 1.8 Components of Nd:YAG Laser [6]

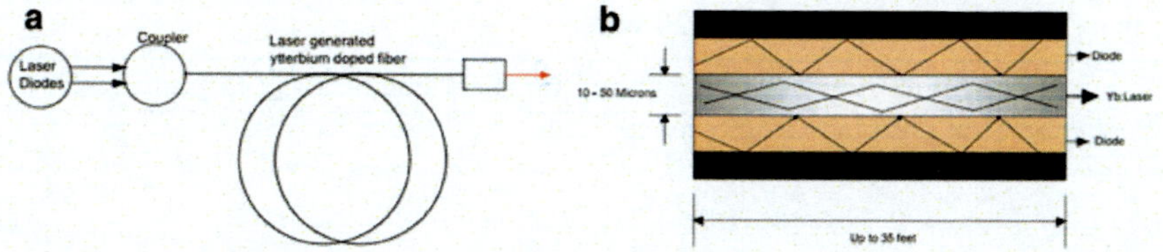

Fig. 1.9 a Fiber laser mechanism [7], **b** fiber

1.5.2 Preferred Materials for Laser Beam Welding

- Material selection depends mainly on thermal diffusivity and the optical characteristics. Selection is not much affected by chemical composition, electrical conductivity, or hardness.
- 300 series stainless steel and carbon steels are the most commonly used materials
- Aluminum alloys and alloy steels should be tested for crack sensitivity. 1XXX series pure aluminum can be welded effectively. Laser welding of Al is difficult because of
  - thermal conductivity,
  - surface reflectivity to infrared and near-infrared laser radiation, and
  - characteristics of the molten alloy including low viscosity and volatility of low boiling point alloying elements like magnesium and zinc.
- Good laser weld can also be obtained for beryllium copper, carbon steel, copper, nickel, phosphor bronze, and titanium.
- Dissimilar metal welding has to be performed after thorough understanding of physical metallurgy of the materials. Precautions are to be adopted to avoid crack propagation by the percentage of alloy elements in the weld.
- Some possible combinations are: Al–Cu, steel–titanium, steel–Cu, copper–phosphor bronze.

1.5.3 Design Considerations

- Laser can be directed, shaped, and focused by reflective optics permitting high spatial freedom in two dimensions. Horizontal welding position is the most suitable.
- Typical joint designs using LBW: lap, butt, and fillet.
- Minimal work holding fixtures required.
- Minimum thickness = 0.1 mm.

- Maximum thickness = 20 mm.
  - Multiple weld runs required on sheet thickness ≥ 13 mm.
  - Dissimilar thicknesses difficult.

1.5.4 Typical Applications

- Structural sections
- Transmission casings
- Hermetic sealing (pressure vessels, pumps)
- Transformer lamination stacks
- Instrumentation devices
- Electronics fabrication
- Medical implants.

1.6 Conspectus of Welding Techniques

This chapter on basic welding terminologies and types of joining processes is concise and ascertains accelerated learning. The applications of welding are eclectic and boundless, and it would be no hyperbole to say that there is no metal industry and no branch of engineering science that evades welding. Welding design of metal components and assemblies is a sine qua non for engineering structures, and it is critical to understand this metal fabrication process. The prime design objectives considered for welding are to manufacture assemblies that perform its intended assignments, warrant facility strategies for fabrication and inspection, enable easy transportation, incur minimum cost, and guarantee safety and reliability. Welding incurs a large percentage of the fabrication cost that comprises the weld procedure framework. Having earmarked considerable expenditure share for welding, imprecise decisions may prove costly which at times can cause catastrophic failures. This is mainly attributed to the multi-variable interdependencies involved in the process, mathematical complexity that defines the process and material parameters expressed in terms of equations and models that lack one size fit solution. To address this critical issue, the next chapter on design importance—for product development—presents contents on case studies exposing the weld failures, factors affecting the weld design, prerequisites for weld, implications of design on weld procedure selection, and design of experiments for optimal parametric window creation. The information presented is substantiated with appropriate literature references.

References

1. America AM (2013) Fundamentals of small parts resistance welding
2. Solid state welding. https://www.mechanicatech.com/Joining/solidstatewelding.html
3. Kopeliovich D, Metal joining technologies, Viewed on 10 May 2020. https://www.substech.com/dokuwiki/doku.php?id=electroslag_welding_esw
4. http://www.weldingcourseinindia.in/WELDER_SKILL_DEVELOPMENT_COURSE_IN_CHENNAI_WELDER_SKILL_DEVELOPMENT_TRAINING_COURSES_IN_CHENNAI_WELDER_SKILL_DEVELOPMENT_TRAINING_AND_CERTIFICATION_IN_CHENNAI.html
5. Arci Welding Industries PTY Ltd., Viewed on 30 Sept 2020. https://arciwelding.com.au/product-category/arc-welding-fundamentals/
6. Dinesh Babu P, Balasubramanian KR, Buvanashekaran G (2011) Laser surface hardening: a review. Int J Surf Sci Eng 5(2–3):131–151
7. Amada Miyachi (2019) Laser welding fundamentals, Amada Weld Tech Inc., Viewed on 13 May 2020. https://dev.amadamiyachi.com/wp-content/uploads/2019/12/Laser-Welding-Fundamentals.pdf

Chapter 2
Decision Making in Welding Design

Abstract This chapter covers the design requirements in the welding processes. The various categories of the welding techniques apply well to different applications and materials involved. Selection of appropriate technique for joining of materials should be based upon the material properties and the application requirements. Implementation of any weld process requires consideration of these factors to determine the optimum process parameters. Quality of the weld connections is determined by the optimum weld geometry, weld process parameters, material process compatibility, and the weld process selection and implementation. This chapter focusses on the significance of design procedures and the factors affecting the design of a weld technique. It also substantiates importance of design with the case studies describing the consequences of improper selection of process parameters and weld technique.

2.1 Introduction

It is important to design, develop, and fabricate optimized cost-effective welded structures accounting for the material, process, and dimensioning methods to achieve the objectives concerning weight, quality, and cost. Therefore, a weld designer should have complete understanding of welding fundamentals, related metallurgy, and fabrication andcharacterization techniques for arriving at efficient design [1–3].

The mechanical properties of the weld depend on the microstructure, chemical composition, and the state of the material which describe whether it is vapor deposited metal, base, or weld metal. Researchers have inferred from experimentations several relationships like embrittlement of grain boundary films, inverse relationship between grain size and strength. They have also developed the cause and effect relationship between input current and heat produced, effect of thermal gradient on the HAZ structure and the variation in structure affecting the mechanical properties of the product.

M. Chaturvedi and S. Arungalai Vendan, *Advanced Welding Techniques*,
https://doi.org/10.1007/978-981-33-6621-3_2

Conventionally, selection of weld process parameters is done on trial and error method based on the discretion of the machine operator. The weld is inspected for the required specification of weld parameters to verify the selection of appropriate input parameters. This process is a time and cost consuming effort toward achieving the desired weldment. This has led to the evolution of several sophisticated techniques and algorithms for parametric optimizations. However, trail and error cannot be completely evaded. This method helps in identifying the limits with which process parameters can vary.

A design engineer has to specify the quality prerequisites in the relevant locations of the structure, as different locations in the structure may be subjected to varying loads due to local stress raisers such as stiffeners, holes, and notches. Inferior quality must be avoided to ensure reliability, safety, and minimize cost. Superior quality than desired on the other hand may result in increased fabrication cost.

2.2 Standards and Specifications

"ISO 3834 Quality Requirements for Welding" was developed by International Standard Organization in association with the International Institute of Welding as International Welding Quality Assurance Standard (ISO 3834). The standard ISO 3834 prescribes the weld requirements for manufacturers to meet which enforce the adoption of good practices [4]. The design is expected to adhere to quality requirement standards.

To obtain consistent weld quality, the welding instructions and elective design protocol are imperative. The weld specification adhered to the welding procedure qualification record (WPQR). On following WPS, achievement of joints with the desired properties will be high.

2.3 Factors Affecting Design

Designers routinely apply the knowledge of the following areas when evaluating the possible techno-economic effects of these on the design of weldments: [2, 3, 5]

1. Physical properties of metals and weldments;
2. Weldability of metals;
3. Welding processes, costs, and variations in welding procedures;
4. Filler metals and properties of weld metals;
5. Thermal effects of welding;
6. Effects of restraint and stress concentrations;
7. Control of distortion;

8. Design for appropriate stiffness or flexibility in welded beams and structural members;
9. Design for torsional resistance;
10. Effects of thermal strains induced by welding in the presence of restraints;
11. Effects of stress induced by welding in combination with design stresses. Design stress is the maximum stress which can be applied to a machine part or a structural member.
12. Practical considerations of welding like the load conditions, use of minimum number of welds, reduction of wastage, plate preparation and direction of rolling on rolled plates, and the selection of proper joint designs for the application.
13. Provide weldment design to the manufactures/operator, including the use of welding symbols;
14. Identification of appropriate welding codes and safety standards.

During the welding, the material is subjected to various forces, and it is essential to determine the impact of these forces on various parts of the weld material. The mathematical analysis of the forces can be done based on the experimental observations and characteristic equations. The sections of the material which will be subjected to these forces should be identified and testing be carried out based on the standards to assure the working in various application environments.

Some of the guidelines that apply for selection of the weld joint based on the application of the weld component have been briefed below [1, 6]:

- Weld metal requirement should be low.
- Preferable to use square grove and partial joint penetration grove.
- Double V or U grove weld is favored for achieving requisite weld metal deposition and minimum distortion on thick plates.
- For accessible weld position, the assembly and the joints should be appropriately planned.

2.4 Expected Weld Characteristics

The following are the characteristics that are expected from any weld:

- The weld dimensions should be as per the specifications required by the application.
- Strength of the weld to match the specified standard
- Joint fit up and the surface finish of the product to fulfill the applicable standards.

The joint efficiency is defined as the ratio of joint strength to the strength of base materials can be expressed as follows [1, 2]:

$$\eta_{\mathrm{joint}} = \frac{\sigma_{\mathrm{joint}}}{\sigma_{\mathrm{bm}}}$$

σ_{joint}—joint strength, σ_{bm}—base material strength.

The weld defects such as pits, undercut, overlap, insufficient reinforcement, surface cracking, bead meandering, remaining groove, and arc strike affect the weld quality and performance.

2.4.1 Example: Design Requirements—For Spot Welding

To understand salient features of design, some of the spot welding design requirements and features are presented below [5, 6]:

- Part thickness: Plates or parts of same thickness result in an even weld nugget. However, there may be mismatch in some cases where in the plate thickness may vary, as illustrated in Fig. 2.1. Standard ratio of plate thickness for effective welds is 3:1 or less.
- Weld proximity: Also described as weld pitch, for obtaining acceptable weld profile and strength, the proximity of the welds should be designed to be ten times the material thickness.
- Edge clearance: Defined as distance between weld spot center to the workpiece end should be maintained as double of the spot diameter. Otherwise, cracks may develop causing a porous weld due to inadequate pressure applied by the electrode. Excessive burrs or metal deformations are also the undesired effects.
- Distance between the form and the weld: In case of metal parts having different formations like a bend or a drawn-in, minimum clearance between the weld spot and the form should be maintained which will help prevent shunting and avoid fabrication defects. The minimum distance should be $(D + R)$ where D—spot weld diameter and R—bend or draw-in radius as shown in Fig. 2.2,
- Spot weld overlap region: Metal pieces that have to be spot welded should be coplanar and thus the weld quality gets significantly affected by the overlap region of both workpieces. Appropriate overlapping lengths of the sheets are to be maintained to attain better dimensional accuracy and to join the sheets

Fig. 2.1 Plates of different thickness [5]

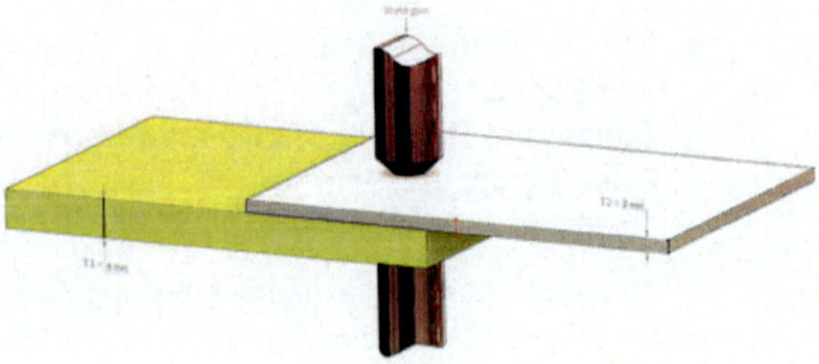

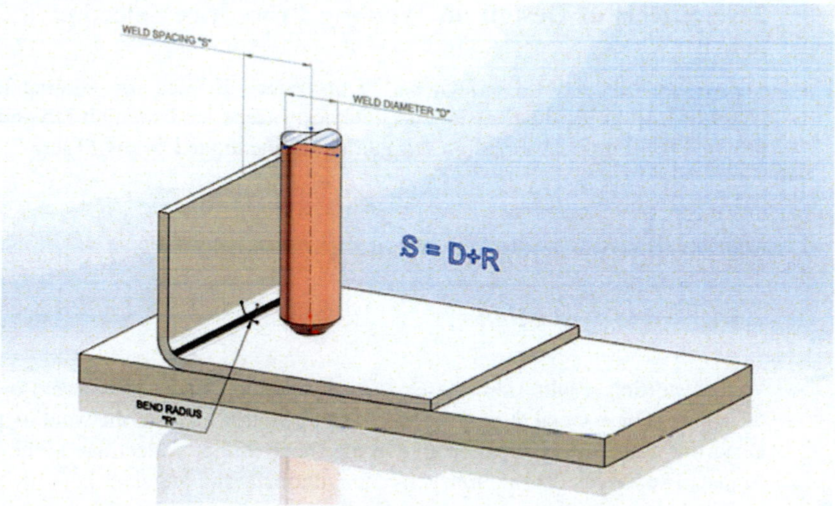

Fig. 2.2 Schematic showing distance of form to the center of spot weld [5]

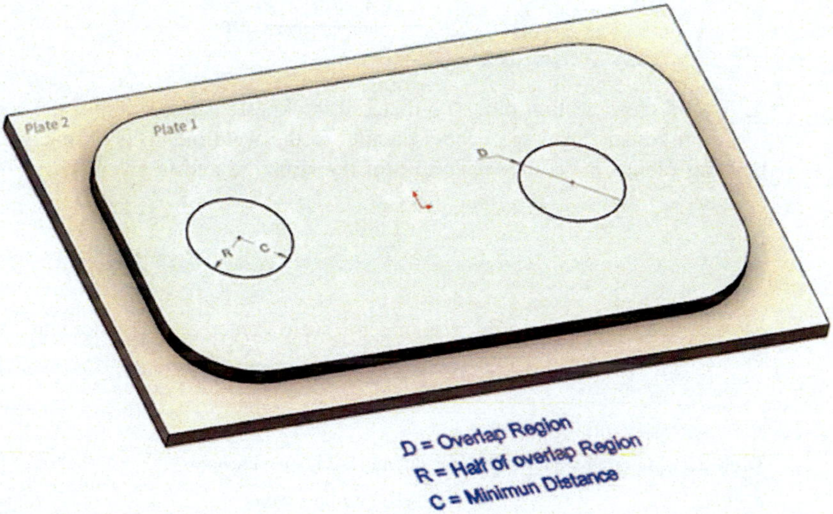

Fig. 2.3 Schematic for overlap region in spot weld [5]

without large deformation as shown in Fig. 2.3. The material being welded and its thickness would decide the overlap length which is usually less than 8 mm in diameter.

2.5 Effects of Design on Welding Procedure Selection

The design of the weld influences the procedure adopted for welding [5]. The factors which guide the decision are thickness, shape, and strength requirement of the final weld product based on the guidelines mentioned below (Table 2.1).

2.6 Weld Distortion and Defects

2.6.1 Weld Distortion

Weld distortion results from the expansion and contraction of the weld metal and the adjacent base metal during the heating and cooling cycle of the welding process. Distortion causes permanent change in the shape of the product due to the induced strains. The strains can be transverse, longitudinal, and angular.

Transverse shrinkage occurs perpendicular to the weld and is caused by the rotational movement in the unwelded portion of the structure.

Transverse shrinkage is given by Eq. 2.1

$$S = \frac{5.16A_w + 1.27d}{t}\,\text{mm} \tag{2.1}$$

A_w—weld cross section mm^2, t—material thickness, d—opening of the root mm.

Longitudinal shrinkage occurs parallel to the weldline. This shrinkage induces bending moments causing distortion of the structure and is given by Eq. 2.2

$$S_L = \frac{0.005L^2A_wd}{I}\,\text{mm} \tag{2.2}$$

d—distance between neutral axis and the weld center line, I—moment of inertia, A_w—weld cross section mm^2, L—continuous weld length.

Table 2.1 Preferred weld method for different weld specifications

Criteria	Preferred procedure
• Thick sections	• Less restraint procedure
• Thin sections	• Weld from one side
• To eliminate preheat requirements	• Low hydrogen electrode process
• For box welds	• Sequencing of fit up, fixturing, and welding is important
• Strong joint	• Weld process giving crown weld

Angular distortion occurs due to the rotation around the weld center line, and the effective shrinkage as a consequence is given by Eq. 2.3

$$S = 0.02W \times \frac{a^{1.3}}{t^2} \text{mm} \tag{2.3}$$

W—flange length, *a*—weld size mm, *t*—plate thickness.

Buckling—This is a deformation observed in thin samples and is caused due to non-uniform heating and cooling.

The distortions mentioned above can be minimized either in the design and the fabrication stage or during the post weld processes.

In the design stage, the following considerations help in minimizing the distortions:

- Design involving minimum number of joints to reduce the amount of weld deposition and distortion
- Location of the welds can be decided to be around the neutral axis
- Selection of the type of joints for example, groove filets instead of plain filets, use of unsymmetrical double V or double U joints
- Weld process should provide rigidity against bending loads.

2.6.2 Weld Defects

Defects in the weld process cover various abnormalities like discontinuity, inhomogeneity, or material structure variation. These abnormalities also occur due to lack of design methodologies, inconsistent processing, and service conditions or some specific material properties which may be classified as under based on their dependency on the weld design or the material properties:

Defects with strong dependence on the weld design:

- Craters—Circular surface cavity that extends into the weld metal. This is caused due to abrupt interruption in the weld arc or sudden variation in the arc travel speed.
- Lack of fusion—Occurs due to non-optimum process parameters of weld speed, weld current, or improper join fit up. The partial fusion may occur between the base metal and the weld metal or between the weld passes.
- Incomplete penetration—Formation of a notch like defect at the opening of the weld root due to incomplete melting. Proper joint design, fit up, and correct choice of electrode for root weld can help avoid this defect. This is caused due to low input current, high weld speed, and improper parametric value selection based on trial and error methods.

- Excess penetration—Involves excessive melting and undesirable protrusion at the root, which may obstruct the fluid flow in the weld applications. This is due to reduction in effective tube diameter.
- Arc strikes—Localized area of material hardening caused due to stray arc melting and may be caused due to scratching of live electrode on the metal surface for striking the arc.
- Weld spatter—During the weld deposition, spherical particles of metal get ejected from the molten pool and get deposited on the base metal surface. Excess spatter around the weld may mask the quality of the weld X-ray picture, thus causing hinderance in the NDT techniques.
- Excess reinforcement—This is the situation when the weld metal deposition exceeds the required deposition. The reinforcement angle affects the fatigue strength, and procedures are to be adopted to maintain them within the limits.
- Root Concavity—Weld root forms a concave shape and causes a suckback of the weld deposit due to excessive shrinkage, high current, slow weld speed, or improper gap in the joint or insufficient filler material.
- Undercut—This is a groove in the base metal left unfilled by the weld metals, thus forming a cavity which weakens the strength of the weld zone as the effective thickness gets reduced. This defect is attributed to the non-optimum input current value. Proper design of the weld parameters with control over the electrode movement can minimize the occurrence of undercuts.
- Overlap—This phenomena occurs when excess of the unfused weld metal extends over the fusion limits and overflows over the surface of the base metal. Improper weld parameters cause this defect, and this overflow can be removed by grinding the weld product.

Defects with strong dependence on properties and impurities in the material or shielding gas:

- Cracks—Metallurgical defects are caused due to thermal gradients or during the grinding process. The cracks cause linear rupture of the material under stress in the weld process. Presence of high percentage of carbon and sulfur and restraint on the joints causes hot cracking.
- Cracks may also occur after the weld process, and they are termed as cold cracks or hydrogen-induced cracks. These are caused due to residual stresses induced in the weld process.
- Slag inclusions—Oxides and non-metallic solid materials in the form of protective slag over the weld metal get entrapped in the weld metal or between the two workpieces.
- Porosity—Caused due to entrapment of gases formed by chemical reactions in the weld. Porosity may occur in the form of clusters or worm holes or as elongated gas hole. This may be caused due to improper weld speed, low heat input, moisture content in the shielding gas, gas flow rate or in correct weld positions indicating requirement of following design methodologies.

In spite of appropriate design, the weld defects occurring in the bead due to impurities in materials will also impact the build quality and the strength.

Thus, for the fabrication of efficient welds in terms of build geometry, microstructure and build properties as per the application requirement, selection of material and processes suitable for the application should be cautiously done in the design phase.

2.7 Design Methodologies

Selection of optimum process parameters is critical for welding domain to obtain cost effective and consistent weld qualities for every weld run. This requires developing a design model to obtain the parametric window for each weld technique. Some of the methodologies used in the welding industry are briefed here and are elaborated in the following chapters based on their implementation.

2.7.1 Energy-Based Model

Computerized simulation and modeling of any welding process can be used to reduce the cost of experiments. This study can help to arrive at a relationship between the weld process parameters and weld geometry. Detailed information related to the different weld process properties like the weld pool geometry or the HAZ or the weld nugget microstructure or the electromagnetic field effect forms the basis of the simulation and the process modeling. The models developed may vary depending upon the consideration of heat source as two dimensional or three dimensional, also at times accounting for the fluid dynamics [7].

The HAZ profile in the base metal and the thermal gradients in the microstructure are also used as the factors to develop simple mathematical models. These models can further be used to estimate the appropriate heat source and the material suitability for the weld process.

2.7.2 Energy Density and Line Energy Estimation

Scientific studies are performed to understand the interactions between weld source and the material. The knowledge obtained can be used to examine the effects of weld process parameters and the energy density or the line energy on the weld bead geometry and the mechanical properties [7].

2.7.3 Full Factorial Design and Response Surface Methodology Used as a Design of Experiments Approach

Various optimization techniques can be applied to develop mathematical models specifying the relationship between input and output variables to achieve the desired weld specifications. Several DOE techniques are in use in the industry to obtain the mathematical models. Some of these are response surface methodology (RSM), Taguchi method, and full factorial design (FFD) [7].

This methodology provides an optimum tool to design and analyze the experiments by eliminating redundant observations and reducing the time, resources, and incurred cost to conduct experiments. DOE uses smart positioning of points in space showing a distribution of input parameter configurations and thus identifying the statistical importance of each factor.

Attributing to various interrelating factors involved in welding, ranging from applied pressure to operating temperature and material properties, DOE techniques are found useful. As an example, for linear friction welding, it was found that the frequency of oscillation, power input, and forging pressure are statistically insignificant for the range of friction pressures studied [4].

2.8 Design Failures—Case Study

Improper design can lead to catastrophic failures in the machine parts and structures. Failures may occur when a part of a system does not perform as per expected behavior. Design failures commonly arise due to misrepresentation or non-consideration of some critical principles or procedures. These failures may also be classified as under, with respect to the components and processes involved [8]:

- Geometrical design
- Identification of optimum process parameters
- Compatibility assessment of weld material and weld technique
- Practical limitations in the weld process.

Failure of a part in service may be either due to inferior design or poor manufacturing of the part for the circumstance in which it is to be used—either stand-alone or in combination.

Below presented are some of the cases from history in which the cause of failure was elated to the weld design.

2.8.1 Case History 1: Magnetic Pulse Welding Failure

MPW is being progressively optimized for effective industrial implementation [9], even including robot arms to effectively handle the portability of the unit in industrial welding cases. Formation of intermetallic compounds as part of MPW process hinders the formation of permanent bonding of dissimilar metals. Transverse cracks are observed across the thickness on the surface due to the shrinkage stresses during solidification. Propagation of such a macrocrack within the intermetallic zone may cause the interface to completely break. This can be avoided by using a design methodology to use critical impact energy criterion to minimize the thickness of the intermetallic layer.

2.8.2 Case History 2: Friction Stir Welding Failure

FSW is used for joining AA2139 which makes up for the key components such as monolithic lightweight armor and vehicle frame structure in military tactical and battlefield vehicles which enhances ballistic limits [10]. The ballistic limit refers to the projectile incident velocity. The FSW joint created in the armor should be able to sustain the blast/ballistic impact. The main failure modes for the FSW joint in the armor were seen as front/back face petaling, ductile hole enlargement, plugging, or spalling. These failures are attributed to high radial and circumferential tensile stresses. As the HAZ/TMAZ material hardness increases, the back face petaling becomes less pronounced and is replaced by spalling.

2.8.3 Case History 3: Laser Welding Failure

Laser weld in the rear axle of a Scania lorry [11]

Rear axle housings were welded to axle ends, and the housings were coated with corrosion protection oil.

The weld was found to have cracks and pores, and also a bulging mid section was observed. Several experiments were done to investigate the cause of defective welds. Initially, the problem was perceived to be due to different materials used in the housing or the material weldability, but with no concrete evidences. Further investigation found that corrosion protection coating would have affected the weld quality. The corrosion protective oil used was DINOL Dinitrol 40. It was revealed that low vaporization temperature of calcium, close to melting point for steel, causes it to vaporize inside the keyhole weld. The axle ends and the housing being a joint without any groove prevent the gas from escaping which led to higher keyhole pressure, thus enlarging it. The enlarged keyhole causes the undercuts and burn through defects.

2.8.4 *Case History 4: TIG Welding Failure*

Failure in Steel fabrication Industry

A thorough thickness crack, as shown in Fig. 2.4 [12], was found in a circumferential weld between a carbon steel pipe fitting and the 439 stainless steel (SS) oil tank panel when heated to high temperature. The consumable specified for manual gas tungsten arc welding (GTAW) was a 309L SS rod using argon shielding gas. The cracked portion of the joint was extracted and opened to reveal the crack surface. From the discoloration of the surface, it was determined that the crack initiated at the location where the pipe fitting met the seam weld of the SS tank. This examination also revealed that complete joint penetration was not achieved.

SEM energy-dispersive X-ray spectroscopy revealed that the weld contained a significant amount of oxygen. A layer of chromium oxide residue was also found on the joint groove surface, indicating contamination due to improper cleaning of the joint after plasma cutting prior to welding.

It was concluded that chromium oxide contamination from the upstream plasma cutting operation initiated the crack under a local stress concentration. Secondary cracking was caused by high residual stresses in the austenite/ferrite duplex matrix.

2.8.5 *Case History 5: Bridge Failure*

A plate section got removed from a bridge in the construction process [8]. To replace this, a joint was intended to be made as a fully penetrated butt weld. But the resulting joint had a significant unpenetrated area in the center of the welded section. The flat space between the two sections was too large for the specified manual welding procedure to penetrate consistently. Thus, it was concluded that the fundamental design of the joint was improper. Partially penetrated welds are permitted in bridge construction under certain conditions when the stress is parallel to

Fig. 2.4 Part of lorry having weld defect [11]

the unpenetrated region. But in this particular case, the sections of the weld were transverse to the fatigue stresses in the structure and the fatigue cracking initiated from this location.

2.8.6 Case History 6: Chernobyl Reactor Accident

Chernobyl reactor accident is quoted to be the most severe accident in the history of nuclear energy. In this accident, the reactor of the fourth unit of the Chernobyl reactor was fully destroyed [13]. This caused a release of a very high amount of radioactive species into the environment. The accident occurred during a turbo-generator test carried out at the chance of the shutdown of the unit for a planned maintenance. The destruction of the reactor happened 6–7 s after the operator pressed the scram button, AZ-5 to insert all control rods into the core.

The state committee for Atomic Safety Survey of the USSR concluded that the main reasons of the Chernobyl accident were serious shortcomings in the design of the Chernobyl reactor as well as inadequate documents regulating a safe operation of the reactor.

It was also observed that there was under design of the absorbers of the channel reactor. These absorbers had special graphite displacers of length 4.5 m. By withdrawal of the absorbers up to their extreme top position above the core, the midpoint of each displacer would be at the midpoint of the core. The length (4.5 m) of the absorbers being less than the height of the core (7 m), water columns of the height of 1.25 m were formed below and above the displacers. Arrangement of the core of Chernobyl Unit 4 is shown in Fig. 2.5.

On moving down of absorbers into the core, their displacers would displace water columns from the lower part of the core. Thus, inserting of absorbers from

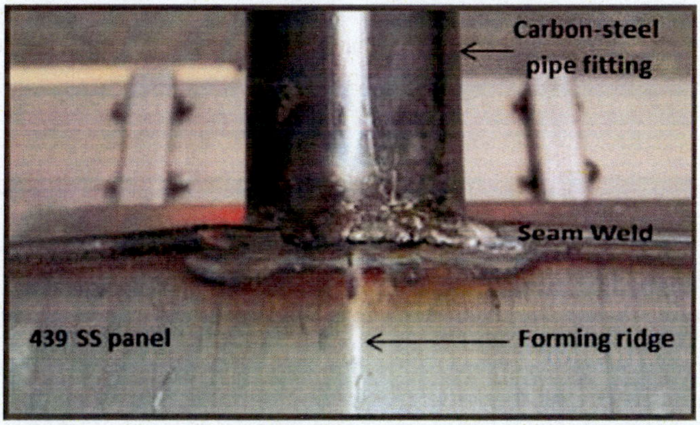

Fig. 2.5 Weld joint between carbon steel piping and oil tank [12]

their extreme top position introduces a positive reactivity into the core because graphite absorbs neutrons much less than water. Specialists named the positive reactivity surge as the "end-rods effect".

This feature occurred occasionally and only by some neutron distributions in the core. Lack of complete understanding of this was evident in a design document where it was told that the positive reactivity surge could appear only in case of neutron field disturbed downward. This statement was wrong. It is known that before pressing the button AZ-5 the neutron field was distorted upward and not downward.

The commission setup for fault analysis stated that severe shortage in the design of the reactor and freak infringements of safety regulations in the construction of the unit 4 are real reasons of the Chernobyl accident.

2.8.7 *Case History 8: Aircraft Engine Failure*

The fan section of a turbine aircraft engine contains fan blade separators known as annulus fillers as shown in Fig. 2.6 [14]. The material specification of an annulus filler is usually 7075 T56 aluminum alloy with a painted coating. Annulus fillers are inserted into the rotor disk (Fig. 2.7).

A B757-200 experienced failure of its right engine during takeoff. Inspection revealed that one of the fan section annulus fillers had detached into the fan case area causing the right engine failure. An ultrasonic inspection six months earlier had not detected cracks on the fillers.

The fault analysis found low cycle fatigue as the failure mode in combination with tensile overloading occurring during the intergranular propagation of the crack. The failed annulus filler contained a cup pattern fatigue crack and an atypical crack. Such a crack can only be detected at a 45° angle. Also, the loading conditions on the annulus filler were not well understood by part designers.

The concept of loading was not completely considered during design. Apart from the centripetal forces associated with the engine acting on the filler during takeoff and anticipated downward loading, an additional loading condition was present—which was ignored. Dynamic modeling could have discovered this design error.

This failure analysis shows the importance of integrating modeling and stress analysis in component design long before the component is put into service.

2.9 Conspectus of Weld Design Studies

To accentuate the far-ranging responsibility which the weld design engineer has, an attempt is made to present the cornerstones of design importance in material joining in this chapter. Starting with prerequisites for weld, this chapter progressively

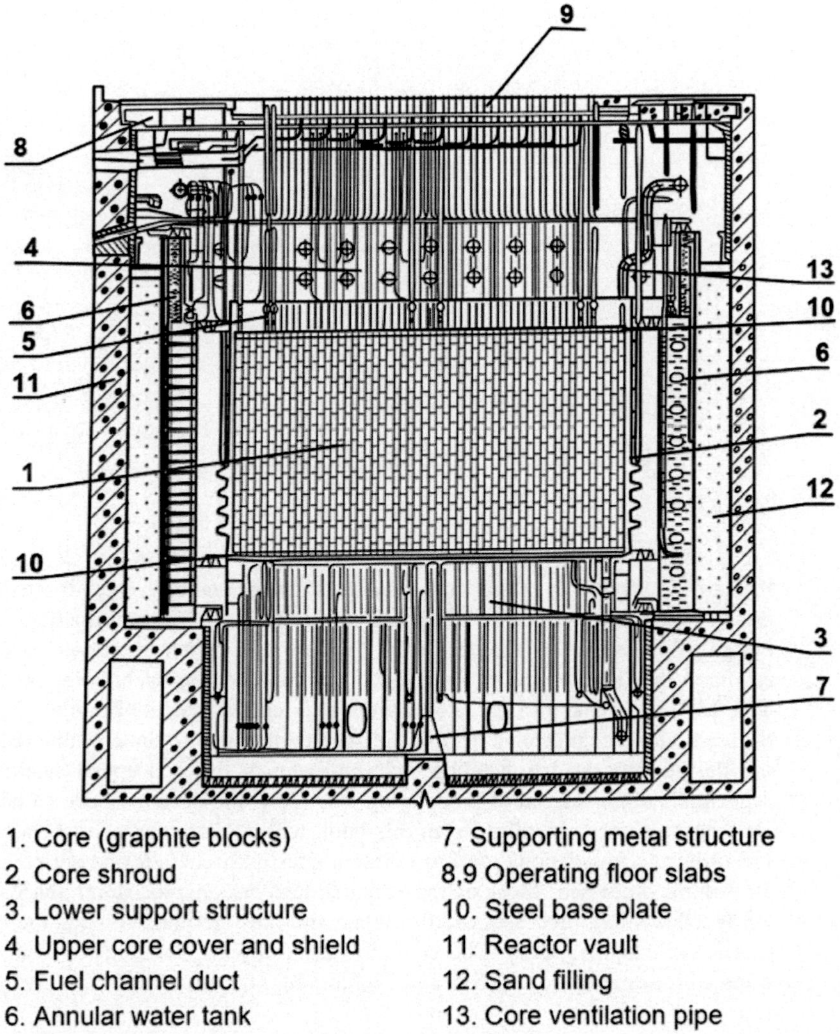

1. Core (graphite blocks)
2. Core shroud
3. Lower support structure
4. Upper core cover and shield
5. Fuel channel duct
6. Annular water tank
7. Supporting metal structure
8,9 Operating floor slabs
10. Steel base plate
11. Reactor vault
12. Sand filling
13. Core ventilation pipe

Fig. 2.6 Arrangement of the core of Chernobyl unit 4 [13]

introduces the factors considered in design of welds and their implications, parametric window creations, and process and procedure selections. Rudimentary education on materials and techniques used by design engineers is purveyed while also pointing out that many of the data and design methods adopted rely on approximations that are a serious impediment to the scientific approaches and their validations. Several frameworks and rules have been formulated on empirical basis which are legit within certain boundaries. There are a few other procedures reported

Fig. 2.7 Annulurs in an aircraft engine [14]

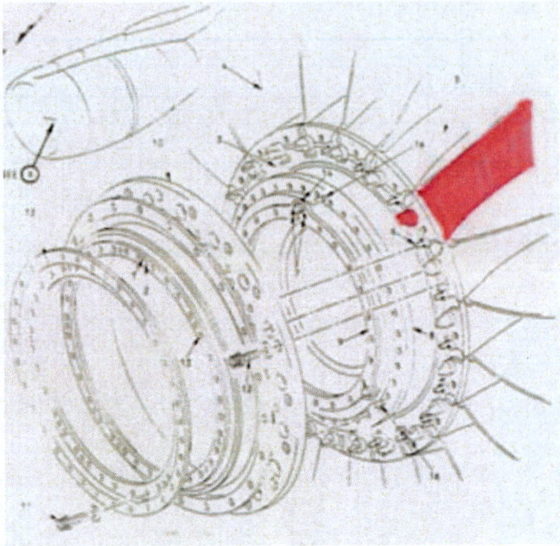

that are applicable to all-inclusive systems whose validity extends beyond the established confines and are discussed in this chapter. The information on weld design furnished in this chapter forms the matrix on which the following chapters reinforce the design aspects involved in various welding techniques, viz. RSW, TIG, LBW, FSW, and MPW chosen from different categories of joining processes. Strong basics recognize and resolve the uncertainty that is often encountered during the design process for welding that undermines the conceptual learning and sequential administration of procedures. Therefore, the design aspects of advanced welding processes are unfolded in this book with a conscious effort being laid on imparting the foundational design terminologies and schemes that are common to all joining processes. Most of the sophisticated design procedures discussed for advanced welding processes are formulaic and have germinated from the conventional and extant process. Consequently, the following two chapters adduce the same with examples from RSW and TIG joining processes that will enliven design learning for advanced join methods.

These following chapters provide a brief history on each of the chosen processes, and the physics underlying them facilitates to put things in perspective and usually helps in understanding concepts which otherwise are difficult to grasp. Besides, the following chapters sequentially present the design procedures discussed in this chapter being extrapolated for real-time industrial welding applications for the processes mentioned earlier. Sample data, process parameters, and the procedures for calculations are illustrated for the processes with appropriate case studies. Formulae for various parameters of welding appear theoretical, whereas in fact they are empirically deduced from a large number of experimental trials. Adequate understanding of weld design terminologies for the relevant joining process is must to accomplish reliable welds and mitigate the catastrophes.

References

1. Keyence, Welding quality requirements. Viewed on 23 March 2020, https://www.keyence.com/ss/products/measure/welding/trouble/quality.jsp
2. Vairis A, Petousis M (2009) Designing experiments to study welding processes: using the Taguchi method. J Eng Sci Technol Rev 2(1):99–103
3. Ogunbiyi B (1999) The influence of power source type on welding performance and weld quality. Exploit Adv Arc Weld Technol, 55
4. http://www.weldfabtechtimes.com/article/iso-3834-the-quality-requirements-for-welding/
5. https://www.machinedesign.com/mechanical-motion-systems/article/21836735/dfm-for-welding
6. Funderbuck RS (2003) Design for welding. American welding society, Viewed on 20 Feb 2020. http://lecturer.ppns.ac.id/munir/wp-content/uploads/sites/14/2015/09/71005378-Design-for-Welding.pdf
7. Khan MMA (2012) Laser beam welding of stainless steels
8. Somers BR, Pense AW (1994) Welding failure analysis. Mater Charact 33(3):295–309
9. Sapanathan T, Raoelison RN, Buiron N, Rachik M (2016) Magnetic pulse welding: an innovative joining technology for similar and dissimilar metal pairs. IntechOpen, pp 243–273
10. Grujicic M, Pandurangan B, Arakere A, Yen CF, Cheeseman BA (2013) Friction stir weld failure mechanisms in aluminum-armor structures under ballistic impact loading conditions. J Mater Eng Perform 22(1):30–40
11. Repper E, Carsbring A (2017) Defect formation in laser welded steels after use of corrosion protection coating
12. Wang W, Weld failure analysis: a case study. EWI
13. Malko MV (2002) The chernobyl reactor: design features and reasons for accident. In: Recent research activities about the Chernobyl NPP accident in Belarus, Ukraine and Russia, p 11
14. Zamanzadeh M, Larkin E, Gibbon D (2004) A re-examination of failure analysis and root cause determination. Pennsylvania, Pittsburgh

Chapter 3
Resistance Spot Welding and Design

Abstract Resistance spot welding (RSW), a thermoelectric process, is a connatural integrant in sheet manufacturing industries for its ability to engineer reliable electromechanical joints. Absence of consumables, economic viability and short processing time are the insignia of RSW process. This process involves interaction of heat and pressure with various possible combinations of electrode and weld materials to create coalescence of parts with the formation of weld nugget. In this chapter, the mechano-metallurgical principles exercising control over the RSW technique is discussed. The peregrinations of researchers to understand RSW and the prominent features discovered for various materials subjected to this technique of joining are comprehensively presented in the subsection. The joint design considerations are vital aspect of RSW as it governs the quality and overall cost of the weld. Henceforth, the terminal section, the distinctive part of the chapter, elaborates on the interaction of the various phenomena involved in the process and their effects on the design criterion with the help of sample design data, calculations and case studies. This chapter exposes the reader to a detailed analysis of various dynamics that occur during the process. The key focus is laid on the electrode-weld surface material compatibility and thermal gradients that causes variation in the hardness, material resistance at the weld nugget, surface modifications, changes in the applied force or variations in the power supply and control.

3.1 Introduction

Resistance welding is employed for versatile industrial applications. It is one of the primitive but prominently used joining process that combines heat, pressure and time for setting the weld. The resistance to the flow of current offered by the metal generates heat. Total resistance includes the bulk and contact resistance of the workpieces and the electrodes. The joule heating causes localized heating at the interface of the two metals to be joined resulting in melting and followed by coalesce of the parts. The heat generation is affected by the resistances at any point in the circuit.

M. Chaturvedi and S. Arungalai Vendan, *Advanced Welding Techniques*,
https://doi.org/10.1007/978-981-33-6621-3_3

Lap joining of metals using spot welding is performed at specific spots by application of calculated value of electrical current and mechanical pressure by the electrodes. The pressure of the electrode tips on the workpiece holds the parts in close contact during the welding process.

Control mechanisms are required to govern the electrode force, current, and the duration of the weld. The pressure is controlled with the arrangement of compressed air in a cylinder–piston system. Current is controlled with the help of a step-down transformer with the control circuitry on the primary side owing to lower current flow in this side. Duration of the current flow is controlled by the principle of ionization of mercury vapor in the ignitron tubes or by the semiconductor principles in the on-off control of SCRs.

Spot welding is a commonly used joining method owing to its reliability and absence of extra material or filler. This joining process is applicable for workpieces in any position. Besides, there is no panel distortion through the welding, making it one of the most preferred processes. The dimensional accuracy of the workpiece is retained during welding with local heating.

The three key stages involved in spot welding process are as follows:

- Squeezing—involves pressing of the electrodes on the metal surface with a defined pressure using a cylinder and piston arrangement.
- Welding—The current from the electrodes is applied briefly through a transformer. The time is controlled either with ignitron tubes or with SCR control.
- Holding—The current is then stopped, but the electrodes remain unretracted for the material to cool via the water quills in the electrode cavity. Figure 3.1 shows the schematic of the RSW circuit.

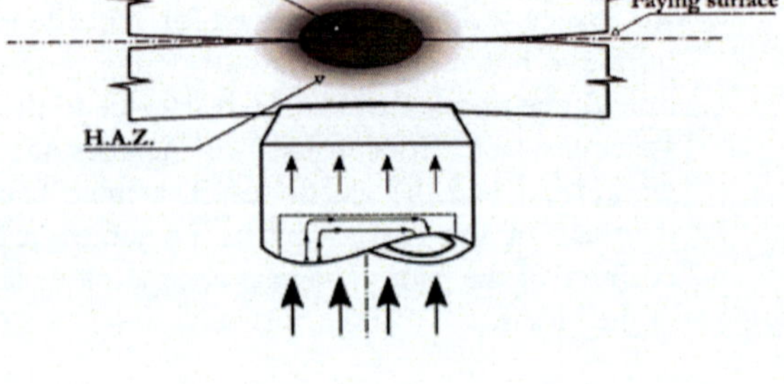

Fig. 3.1 Schematic of RSW circuit [1]

Weld time typically ranges between 0.01 s and 0.63 s, respectively, and is governed by the diameter of the electrodes, specimen thickness, and the electrode force [1].

Weld Nugget

The melted and solidified region of the base metals is referred to as the weld nugget. The nugget formed is classified into three zones known as

- Fusion Zone—FZ
- Heat Affected Zone—HAZ
- Base Metals—BM

The heat generated in the fusion zone is dissipated by thermal conduction through the base metal, resulting in the formation of the heat affected zone. The remaining—Base Metal remains unaffected by the heat. Figure 3.2 shows the cross section of a nugget [2].

The size of the weld nugget is majorly dictated by the welding time, welding current, electrode tip diameter and electrode pressing force. These are the four main control parameters that enable a weld nugget to be formed that provides adequate joint strength for the planned application [3]. It is difficult to characterize the spot weld joint for its strength due to the limitations of nugget area measurement methods. The bearing capacity is usually used to express the characteristics of the welded joints as this is the parameter that describes the ability of spot weld joint to continue load between joined parts [4]. The experimental results indicate that shear load bearing capacity is controlled by the welding voltage during tensile–shear static test [4] as observed in Fig. 3.3.

The welding variables range for which permissible spot welds are achieved for a prescribed material joined with a defined value of electrode force is shown as a graphical representation often termed "spot weld lobe curve" [5] in Fig. 3.4.

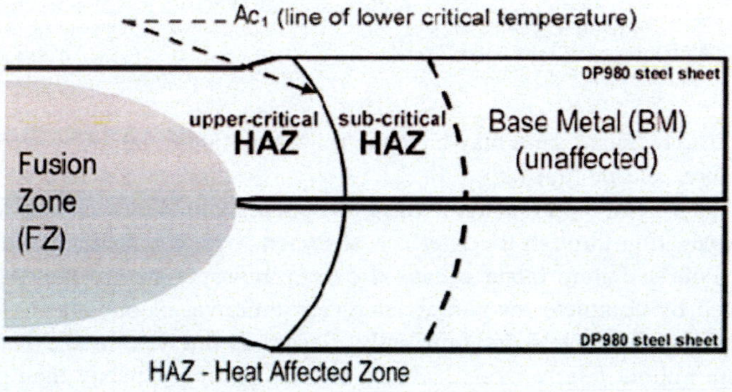

Fig. 3.2 Cross section of RSW nugget [2]

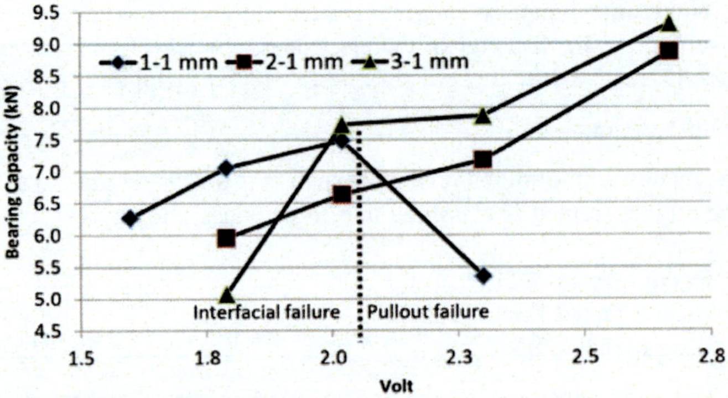

Fig. 3.3 Effect of voltage on load bearing capacity [4]

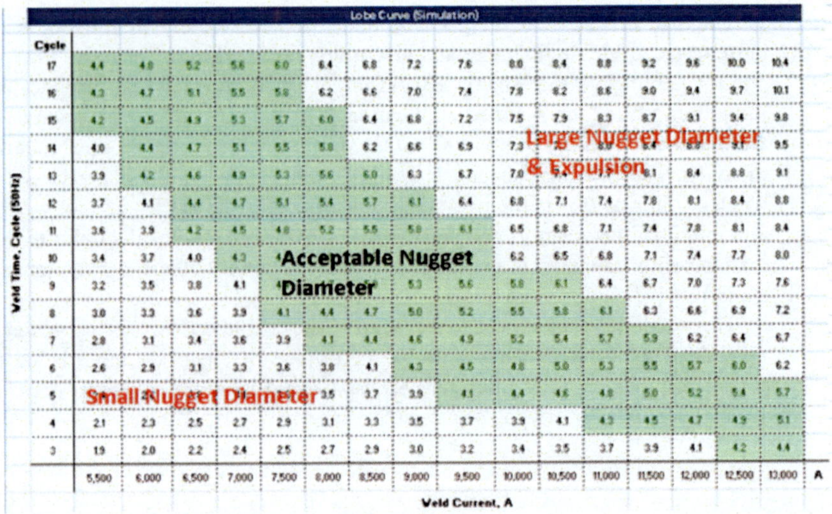

Fig. 3.4 Spot weld lobe curve [5]

The resultant weld may have the following failure modes: interface failure, weld failure, and pullout failure [4, 6].

Interfacial failure is due to lack of bonding and is indicated in the form of crack propagation through the interface of joined sheet and fusion zone (weld nugget).

Pullout failure mode is caused due to improper size of the nugget and is indicated by complete (or partial) nugget withdrawal of one sheet. The load bearing capacity of the weld is significantly affected by the weld mode. Those characterized with pullout failure mode had higher load bearing capacity than interfacial failure mode samples [4]. Thus, to ensure weld reliability in the product, pullout failure mode should be the acceptability criteria.

3.2 Process Applications

RSW is predominantly used in the electric appliances and components of automobiles, aerospace, rail, refrigerators, washing machines, metal furniture, electronics, construction and aviation industries. This joining process is often employed in industries demanding higher volume and rate of production, respectively, and is predominantly adopted for steel car fabrications since early twentieth century.

The automotive industry is the major user of RSW due to its low cost, high speed, simple mechanism and applicability for automation. RSW is used for joining auto body assembly, with about two to six thousand spot welds performed on each manufactured car.

3.3 Compatible Materials

Steel and specifically low-carbon steel have low thermal conductivity and higher electrical resistance and thus are the preferred material for spot welding. Zinc-coated galvanized steel requires higher welding currents to weld than uncoated steels.

Copper and its alloys can also be joined by RSW; however, having same material as workpiece and the electrode may pose some constraints in the spot weld since the heat generation in the electrodes and workpiece will be similar. Thus, molybdenum and tungsten materials are considered for electrodes in case of spot welding of copper due to the high electrical resistance and melting point of these materials compared to copper.

Other materials that are commonly welded using this method are austenitic and ferrite grade steels, nickel alloys, and titanium.

3.4 Fundamentals of Resistance Spot Weld Process

Weld formation relies upon the appropriate control of electrical and thermal processes involved in welding. Power source is the essential component that supplies appropriate heat to achieve a consistent weld. RSW yields high-quality weld on proper selection of process parameters and the appropriate material. Current level, electrode force, and materials being welded are the variables expected to cause significant variation in the shape of the weld.

In most cases, thousands of amperes is required for making the weld. Such amperage values, flowing through a relatively high resistance, will create enormous heat in a short time. To accomplish efficient resistance spot welds, it is necessary to control the current flow duration.

The time for which the joint is subjected to the current flow is dictated by the composition of the material and its thickness, density of the current flowing, and the cross-sectional area of the contact surfaces of the welding tip.

The resistance of the materials to be joined, the specimen thickness, and desired nugget size will determine the weld voltage.

From the various experimental trials and simulation techniques, it has been observed that weld time, current, and electrode force are the vital factors to obtain a good quality weld. Also, 20% of the weld quality issues are power supply related [6].

These parameters solely depend upon the type of base materials and their thickness.
In an electric circuit, the expression for the generated heat is given by:

Heat Generated in Electric Circuit:

$$Q = \int I^2 R \mathrm{d}t = \int \left(\frac{v^2}{R}\right) \mathrm{d}t = \int vt \times I \mathrm{d}t \tag{3.1}$$

The resistance, the current, and voltage vary with time; thus, the heat expression is written as an integral over time.

The load-dynamic resistance is the key component in this model. The load is the welding machine, and its impedance consists of the following components [6]:

(1) resistance of the electrodes
(2) bulk resistance of the workpieces
(3) contact resistance between the electrode and workpiece
(4) contact resistance between workpieces and
(5) resistance of the cables.

Resistances 1 and 2 are considered as miscellaneous loss resistances.

The resistance of the weld spot (resistance components 2, 3, 4) changes as the workpiece melts as shown in Fig. 3.5, thus affecting the voltage between the electrodes.

Bulk resistance is sensitive to temperature and independent of pressure. For all metals, bulk resistance increases with temperature. Contact resistance is a strong function of pressure and is also affected by the contact surface. This resistance is high at the start of the weld causing initial heat dissipation. Further, the heat and the pressure soften the material at the electrode–metal interface and contact resistance value drops.

The resistance variation in different stages [6] is described below and is shown in the Fig. 3.5

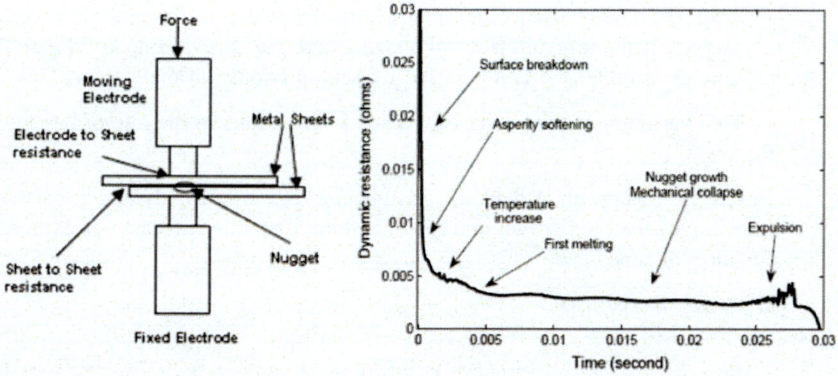

Fig. 3.5 Various resistances and theoretical dynamic resistance curve [6]

- Stage I: When the workpieces are brought into contact under the pressure provided by the electrode force, on application of the supply voltage, current flow at the contact points experiences the resistance between the electrodes. The resistance at this point comprises of:
  - bulk resistance of the two workpieces
  - the two electrode-to-workpiece contact resistance
  - the workpiece-to-workpiece contact resistance.

  Due to the high contact resistance of the workpieces, heat generation will be concentrated at the workpiece surfaces. This heat further causes a drop in resistance as the surface contaminants break down.
- Stage II: Metal-to-metal contact of the workpieces exists after the surface contaminants have got broken down in the stage I. The contact surfaces are uneven and thus have many disjoint contacts. This reduces the contact area to a small fraction of the workpiece surface. This results in a relatively large interface resistance.
  Concentrated heating at the workpiece surfaces causes the temperature to increase and thus the resistivity also increases. Also, due to softening of the asperities, contact area increases and causes the resistance to decrease. The resistance at the workpiece surfaces effectively increases with increase in temperature, surpassing the effect of increasing softened region.
- Stage III: With increase in the temperature of the workpieces, the resistivity increases and is evident from the resistance curve Fig. 3.5.
- Stage IV: The increase in temperature of the workpiece causes an increase in the resistance, but the effect of continued melting causes an increase in the cross-sectional area available for the current flow and also the increased softening, shortens the path for current flow. As a combined effect, the bulk resistance effectively decreases.

- Stage V: The resistance continues to decrease due to the growth of the molten nugget. If the nugget grows to a size where the surrounding solid metal cannot contain it under the compressive forces, an expulsion may occur.

This variation of load resistance leads to variations in the rate of heating during the weld.

Based on experimental data, an analytical expression for dynamic resistance has been established by Brown and Lin [7] where dynamic resistance is expressed as a function of time.

Dynamic Resistance:

$$R_{\text{dynamic}} = (1.1433^t - 0.8867) \cdot [u(t) - u(t-1)] + \left(0.7937^{(t+5.4253)} + 0.03\right) * 0.025[u(t-1) - u(t-20)]; \quad (3.2)$$

R_{dynamic}—dynamic resistance in mΩ, t—time in ms, $u(t)$—unit step function.

3.5 Resistance Spot Welding Machine Details

For an optimum nugget formation, it is important to control the magnitude and duration of the current and the electrode force acting on the workpieces [8].

For a sample of low-carbon steel of thickness 1/16″, the parameters required are 1000 A current for 0.25 s and a force of 600 lb. Transformers are used to supply the high-current requirement for the spot welding. Time control is implemented using either ignitron tube or SCR control. The operating mechanism of both the devices is that a small electrical signal applied to the device allows it to turn on for a fraction of a second and the required current on the primary of the transformer can flow through the device. The device turns off on removal of the electrical signal. The force required to hold the two parts together and press them together to aid the weld is commonly provided by cylinder–piston arrangement with the cylinder attached to the welding machine and the piston to the upper electrode.

Appropriate welding control is required to precisely coordinate the electrical and mechanical operation of the weld machine. The control will provide the control signals for the SCR and for the electrically operated air valve for mechanical control. The ignitron or the SCR circuit provides the switching action and is thus connected in series with the transformer as in Fig. 3.6 on the primary side because of lower current requirement.

The resistance spot welding machines are constructed so that minimum resistance gets included due to the device components which are the transformer, flexible cables, tongs, and electrode tips. These machines are designed to bring the welding current to the weldment in the most efficient manner. The optimum energy

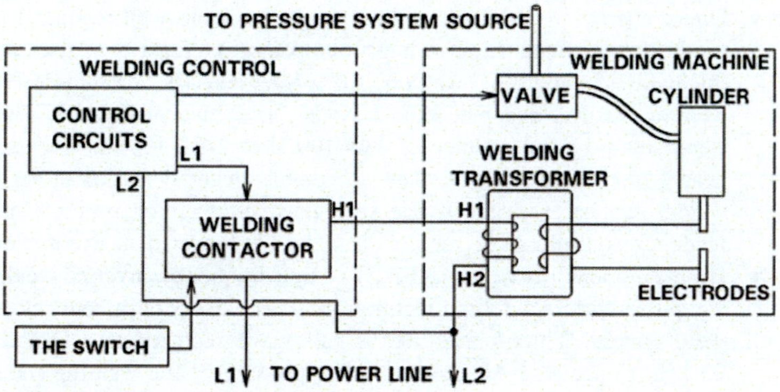

Fig. 3.6 Machine and control link [8]

requirement as per the weld schedule is provided by the regulated power supply. For the power electronics-based supply design, buck converter with PWM control is used and the weld resistance is modeled as a variable resistor. High-current, low-voltage supply is the requirement for RSW. There are various types of power supplies that can be employed for the resistance spot welding [9].

Some of the power supply technologies are described below [9]

- Capacitive discharge (CD welder)—Energy from the power line is stored in welding capacitors. Stored energy is then rapidly discharged through a pulse transformer to produce flow of current through the welding head and the workpieces as illustrated in the Fig. 3.7. Dual pulse feature is used in some weld procedures in which the first pulse is used to displace the surface oxides and plating and the second pulse produces the weld. Single short pulse of duration 1–16 ms causing rapid heat at the welding interface is also utilized for obtaining desired weld. Polarity switching is used when a wide variety of polarity sensitive dissimilar metals are to be welded by the same machine. Length of the output pulse varies for different types of materials.

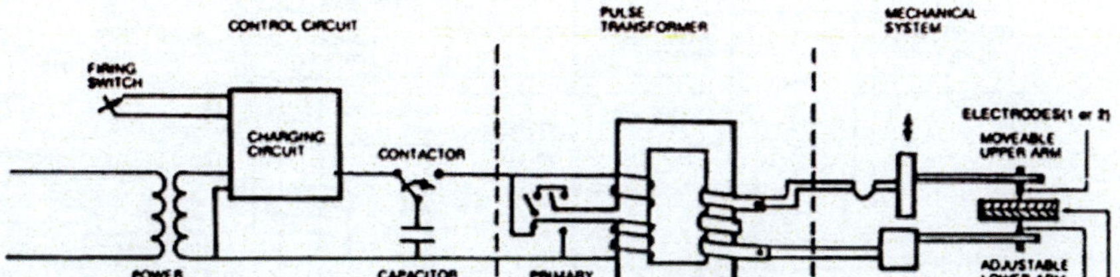

Fig. 3.7 Schematic of a capacitive discharge welder [9]

- Direct energy (AC)—Utilizes the energy from the well-regulated power line is used for the weld. Line voltage compensation feature adjusts for the power fluctuations in some AC welders. AC welder consists of a step-down transformer, welding control, and mechanical system as shown in Fig. 3.8. The transformer steps down the line voltage of 480–100 V to 2–20 V with secondary current of nearly more than 1000 A. The acceptable material thickness for utilizing this welder can be assessed by the secondary short circuit rating. Constant current feedback control can be employed for welds longer than five cycles.
- High-Frequency Inverter (HFDC)—High-frequency inverter welders (Fig. 3.9) use pulse-width switching technology with closed-loop feedback to control the weld energy. Three-phase AC is full wave rectified to DC and switched at 25 kHz to produce AC current at the primary of the welding transformer. The secondary current is then rectified to produce DC welding current with an imposed low-level AC ripple. In this form of power supply, electrodes and the parts positioning can be tested with a preweld check pulse. Also, benefits of reduced power consumption, high-speed feedback circuitry for efficient control and need of smaller transformers are offered by this form of supply.
- Linear DC (Transistor Direct Current)—Transistor welders are also known as linear DC welders. In a transistor welder, energy is stored up in capacitors and released through transistors. This produces a result similar to the high-frequency inverter power supplies. The output is in the form of clean square waves with rapid rise time, also.

  Some transistor welders are closed-loop designs, utilizing transistor-controlled feedback with fast response time, which are able to monitor the condition of the weld using one of three feedback modes, "constant current," "constant voltage," or "constant power." These feedback modes offer great advantages such as a superior process consistency between multiple weld stations, reduced number of bad welds, and real-time automatic adjustment for variations in part thermal loading and plating. Linear DC welders do not use a transformer. These power supplies have the best low energy control, making them the favorite for welding microwires and then foils. The major limitation is that the duty cycle is typically

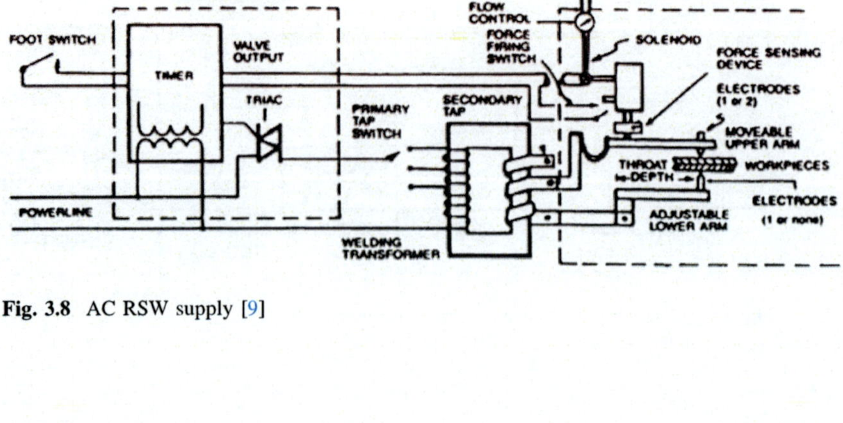

Fig. 3.8 AC RSW supply [9]

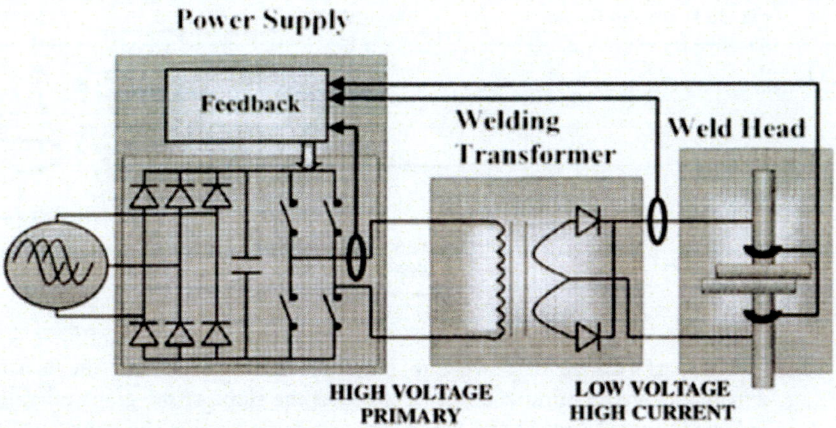

Fig. 3.9 HFDC power supply circuit [9]

much less than one weld per second. Transistor-controlled feedback enables feedback response times of nearly 5 μs. Constant voltage feedback has an ability to prevent arcing due its fast feedback response and also provides optimum weld power distribution based on the part resistance.

The general welding current types applied in RSW are shown in Fig. 3.10 that includes single-phase AC most commonly used in production sectors, three-phase AC or DC with a high frequency, the capacitor discharge (CD), and the DC middle frequency inverter. Welding current value that of the root mean square (RMS) is used in the machine parameter settings and the control mechanism [10].

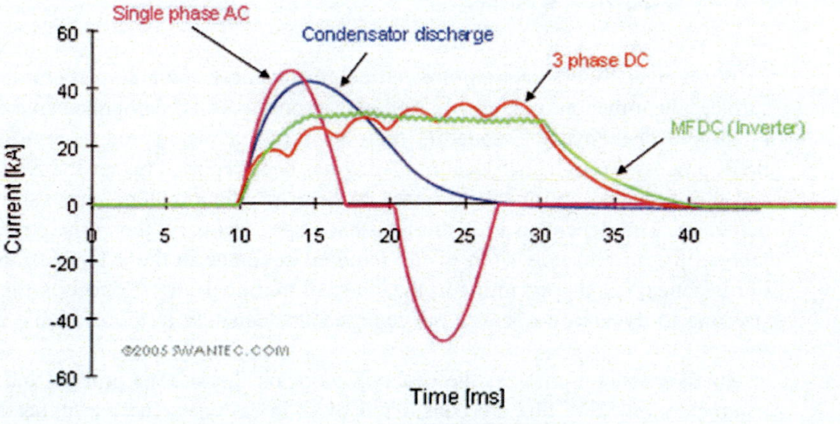

Fig. 3.10 Welding current for different power supply mechanism [10]

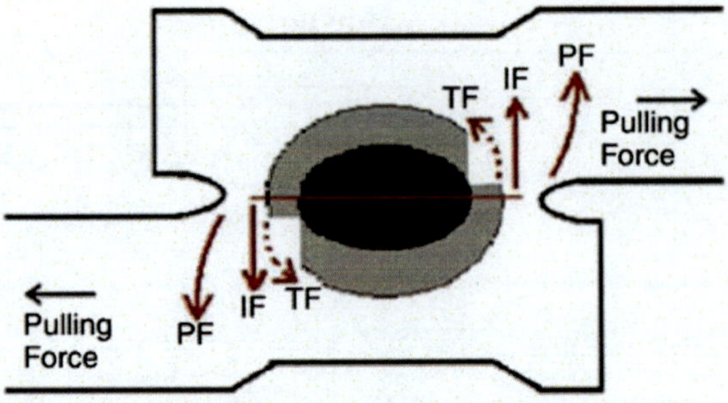

Fig. 3.11 Sites of failure initiation [3]

The performance of a welding power source is dependent on its static and dynamic characteristics which are defined as the slope of the graph relating output voltage to output current and the inductance, respectively. The inductance controls the rate of rise of welding current in response to a change in arc voltage, while the slope limits the amount of the short circuit current attainable. Conventional transformer-based welding power sources are usually designed with fixed slope and inductance levels.

For optimum process behavior, the power source characteristics should be controllable to suit specific welding situations. Inverter-based power sources provide this capability electronically through high-speed control of the welding current, which makes it possible to set a target current and a rate of current change separately, thus achieving conventional slope and inductance control.

Welding power source should have dynamic characteristics to provide the required arc voltage and weld current according to the weld conditions and thus provide arc stability in the weld duration.

3.6 Literature Review

RSW involves the interaction of electrical, mechanical, thermal, metallurgical, and surface phenomenon-making it a complex process. Thus, comprehensive analysis by simple mathematical modeling may be a misleading approach. Simulation of RSW process through analytical modeling has been reported by several researchers, and those studies are mostly directed to heat transfer problems and surface phenomenon. The dependence of the resistance spot welds on the complete range of process parameters falls short of the required coverage in the available literatures. For instance, results pertaining to the study of thermoelectrical effect on the nugget size due to dynamic resistance are reported inadequately in the available research reports.

This section summarizes the research work on the various process and control parameters on RSW that are considered by different researchers to understand the impact on the weld quality.

Charde [3] analyzed the parametric influences accounting for the weld time, current and force variations by maintaining electrode tip diameter constant on the spot weld nugget size. They concluded that the diameter of the weld nugget and the tensile strength of the weld (Fig. 3.12) are proportional to the weld current and weld time, respectively. On the contrary, diameter of the weld nugget is inversely proportional to the electrode force. Besides, an increase in electrode pressure results in decrease of weld diameter, thus increasing the possibility of occurrence of weld failure. Different failure modes were observed while characterizing different weld strengths depending upon the nugget diameter. In the range of 4–7 mm nugget diameter, the failure zones observed were interfacial, partial, and tear failure, and Fig. 3.11 indicates the sites of the initiation of these failures. Moreover, the hardness of weld zones was recorded to be higher than that of the base metal.

Salem [6] investigated a power supply strategy and a generic power control mode to improve the consistency of the spot weld. In the power strategy, hall-effect current sensor and tip voltage sensor were employed to measure the two parameters and use them as feedback to control the PWM DC–DC converter. The signal from the voltage and current measurement circuits on being filtered through a low-pass filter was then used by the DSP board, as well as the PID analog control circuit used in the investigation.

The sensing circuit for the tip voltage is designed using the ZXCT1010 chip, which is a high-side voltage monitor. This clip outputs up to 2.5 V sense voltage. Figure 3.13 shows the circuit schematic for tip voltage measurement.

The hall-effect sensor (Allegro 1302) transforms the magnetic field into a proportional voltage signal with a resolution of 1.3 mV/Gauss in the arrangement as depicted in Fig. 3.14. Through experimental calibration, the scaling of the current sensor is obtained as 4.659 A/mV.

Prashanthkumar et al. [11] carried out process parameter selection for RSW through full factorial design of experiment and thermal analysis using SYSweld to obtain optimum value of current and weld time for 2-mm Cold Rolled Closed

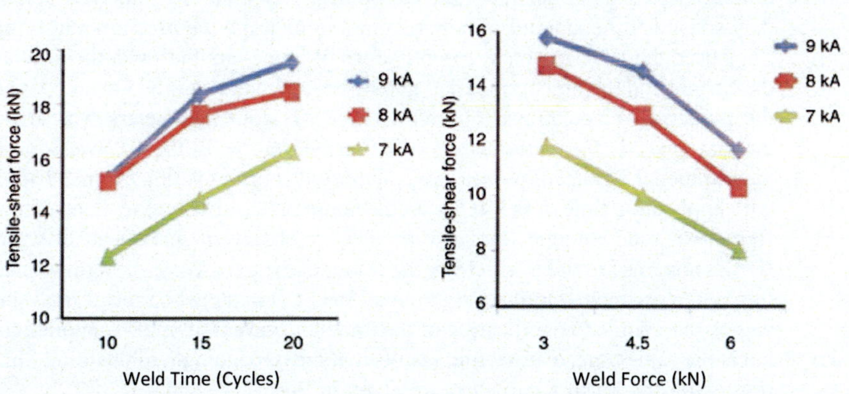

Fig. 3.12 Dependence of tensile–shear force on weld time and welding force [3]

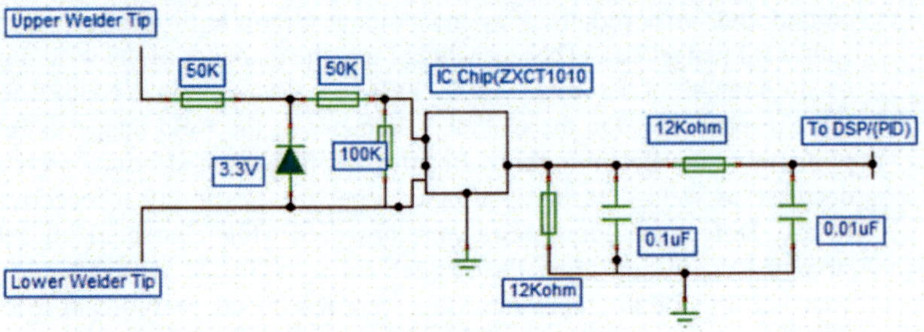

Fig. 3.13 Schematic for electrode tip voltage measurement circuit

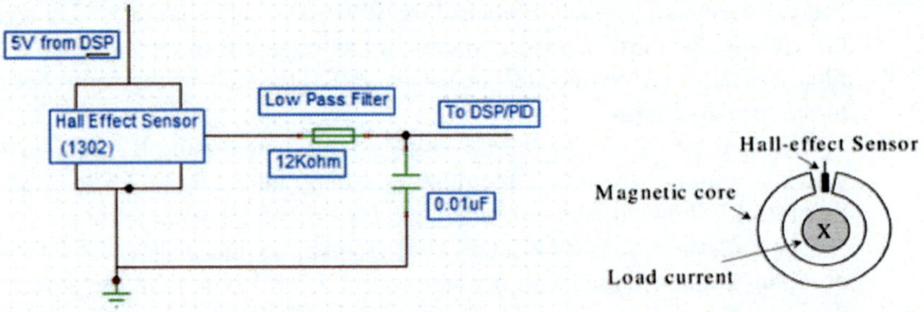

Fig. 3.14 Weld current sensing circuit using hall-effect sensor [6]

Annealing (CRCA) sheets for application in automotive industry. Automotive industries require close control of process parameters to match the requirement of different sheet metal thickness for car bodies. Based on the numerical calculations, they arrived at the amount of heat required to melt the required amount of material. They used the DOE to obtain combinations of input variables with the limitations on the current and the time required to get the desired weld quality. On comparison with the numerical calculations, optimum range of input parameters was arrived and thermal analysis then conducted on the arrived range. In the SYSweld simulation, they studied the effect of process parameters with current varying from 3300–3928 A with application time of 0.3–0.5 s on the temperature distribution, heat affected zone formation, and nugget dimensions. They observed that heat affected zone (HAZ) remains around 5 mm from the center of the axis. Weld strength was analyzed through experimental trails using the pull load test and peel off test. Metal sputtering was observed due to overheating of the joint at a current of 3928 A applied for 0.5 s, but at the same time, overheating results in increased strength of the joint. Simulation results for one set of parameters are given in Figs. 3.15 and 3.16.

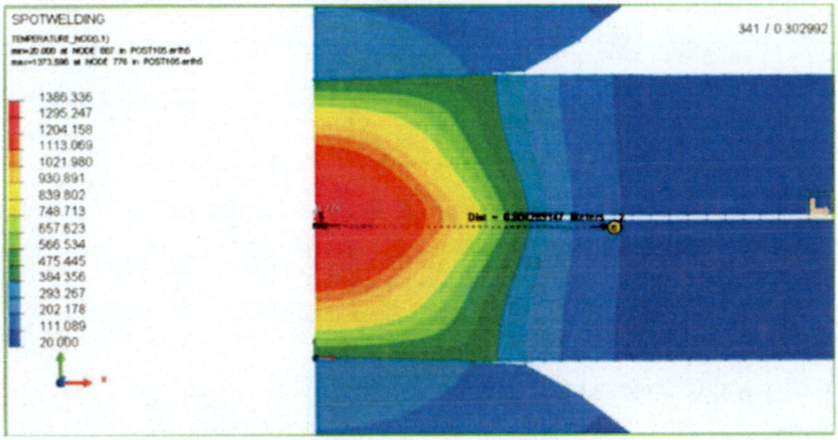

Fig. 3.15 Temperature distribution [11]

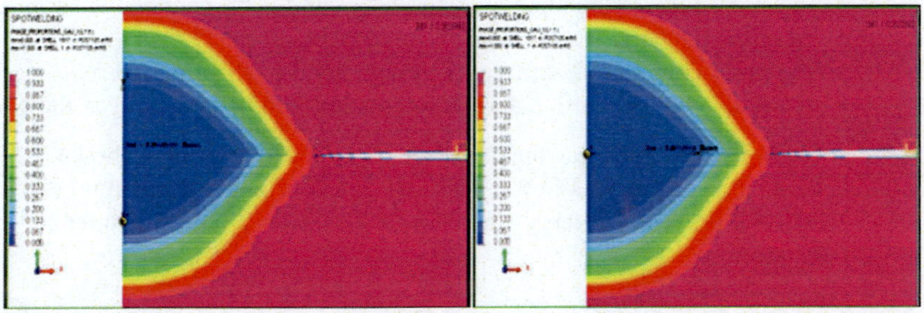

Fig. 3.16 Nugget dimensions for one sample [11]

Triyono et al. [4] studied the failure modes of RSM for thin plates with respect to the weld plate thickness and load voltage to estimate the critical nugget diameter. They found that the experimental nugget diameter decreased if sheet thickness increased [12]. Also, the critical nugget diameter was found to be higher than the nugget diameter obtained by the formula set by American Welding Society.

Bondarenko et al. [13] worked on the power supply design for RSW to provide optimum input current and to increase the energy efficiency of the output pulse generator which provides current pulses to the load.

The input converter is considered to be of energy storage type which comprises of a charger and an energy storage element, i.e., the supercapacitor having maximal voltage of 2.7 V. They suggested multiphase topology of charger with power factor correction based on single-ended primary inductance converter (SEPIC) converters shown in Fig. 3.17. The SEPIC converter has lower input current ripples and better

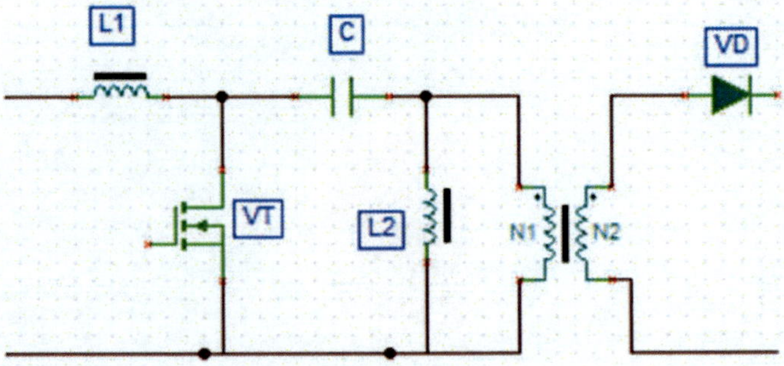

Fig. 3.17 SEPIC converter circuit

controllability than the Cuk converter, and it meets the following requirements to charge the supercapacitor:

- the availability of the input inductor, which forms the continuous current, drawn from the network
- the possibility of soft start and soft regulation of the input current
- the possibility of the transformer integration into the structure of the converter.

They also observed that the use of synchronous transistors as shown in Fig. 3.18 instead of reverse diodes in the buck converters of output generator cells provides substantial increase in energy efficiency of the power supply due to the low on-state resistance.

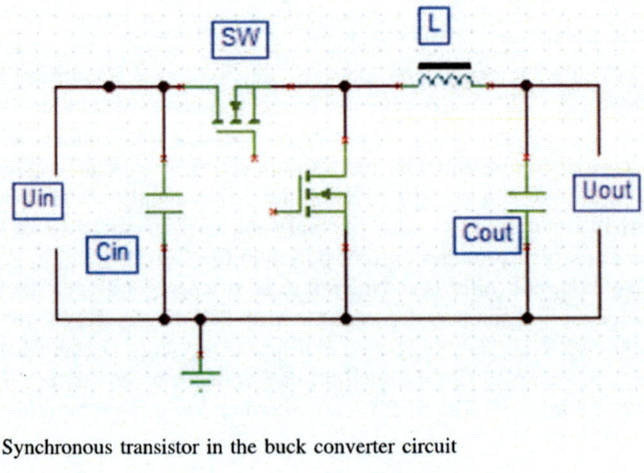

Fig. 3.18 Synchronous transistor in the buck converter circuit

Mehta and Haque [14] worked on integral switching cycle control (ISCC) technique as shown in Fig. 3.19 for RSW power supply. ISCC differs from conventional integral cycle control with respect to the off time. ISCC uses a microcontroller which calculates the approximate number of cycles required for the welding current. They concluded that ISCC has minimal harmonics and proves more reliable compared to phase control switching and conventional on-off control method, but at an increased cost.

Saleem [15] developed a control scheme using DSPIC33FJ16GS502 controller to drive a high-frequency full bridge converter used for the inverter drive in resistance spot welding equipment (Fig. 3.20).

This work has also suggested a protection scheme with the use of high-speed fuses and Typower IGBT fuse, (Fig. 3.21) to protect the power switches (IGBT) used in the converter in case of a circuit failure when the stored energy of the DC link capacitors may rupture the IGBT device. Use of Typower IGBT fuse introduces less inductance in the circuit.

Brown and Lin [7] built a low cost, highly flexible power supply for RSW using constant power control mode to produce consistent sized nuggets. Their designed supply used PWM with low-cost MOSFETs at 20 kHz switching frequency to convert the power of a 12 V battery to the weld current up to 800 A.

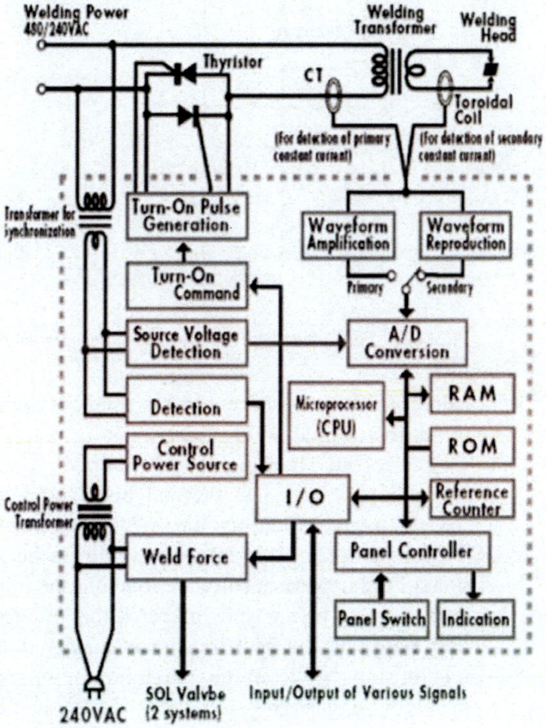

Fig. 3.19 Block diagram for ISCC strategy [14]

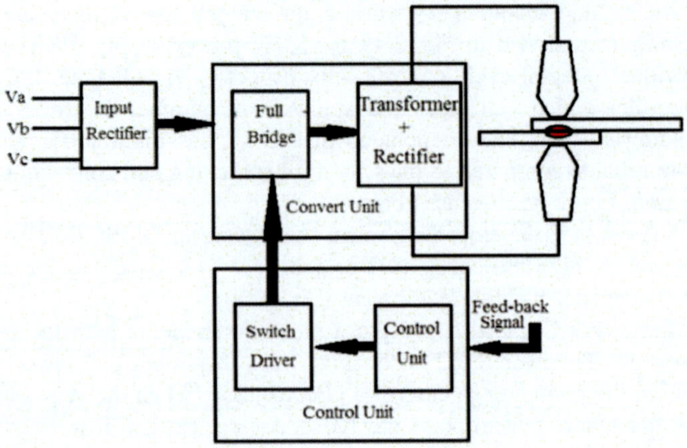

Fig. 3.20 Inverter-type spot welding circuit [15]

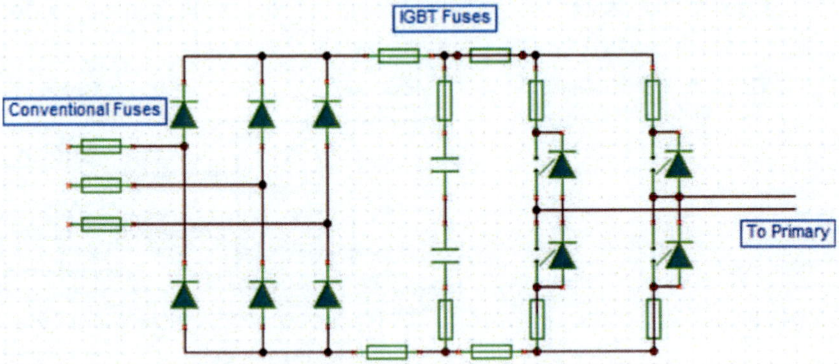

Fig. 3.21 Fuse placement for inverter-based power supply

Microprocessor-based control is used for applying constant power control scheme to get the desired weld quality.

Yeung and Thornton [16] developed a parametric model as shown in Fig. 3.22 to predict the transient thermal behavior of the spot welding electrode cap. The measurements of various parameters are mentioned in Fig. 3.25.

The parametric model uses conjugate heat analysis with the help of computational fluid dynamics software to avoid the uncertainty introduces with heat transfer coefficients. The analysis indicated that convective and radiant heat losses were not important and the heat transfer is mainly due to conduction. They found a linear relationship between the maximum temperature and input power. The overall behavior suggested a slow thermal response but a fast heating rate.

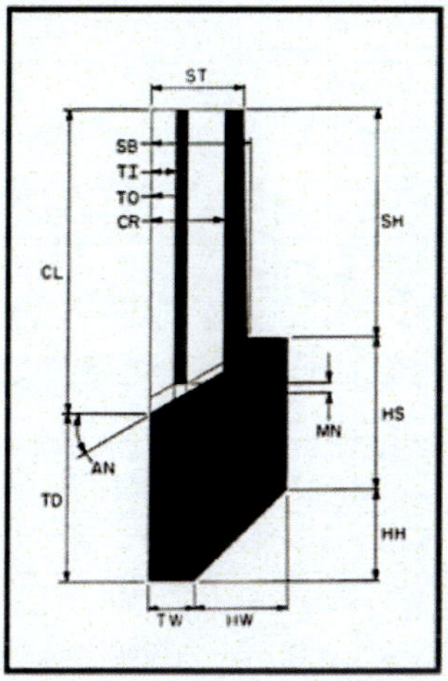

Parameters	Values
AN	30°
CL	22.0 mm
CR	4.8 mm
HH	6.0 mm
HS	10.0 mm
HW	6.0 mm
MN	0.6 mm
SB	6.4 mm
SH	15.0 mm
ST	6.0 mm
TD	9.0 mm
TI	1.6 mm
TO	2.4 mm
TW	3.0 mm

Fig. 3.22 Parametric model of RSW and weld cap geometry parameters [16]

Podržaj and Simončič [17] proposed a controller based on fuzzy logic that has the ability to detect expulsion and halt the joining process on its occurrence. Their proposed algorithm would significantly reduce the thermal stresses on the electrode after expulsion. Applicability of the algorithm depends upon the number of signals required to detect expulsion.

Cho and Cho [18] developed an analytical thermoelectric model to predict the growth of nugget geometry and to analyze the temperature distribution in the weldment. They considered the effect of thermoelectric interaction at the weldment interface on the internal heat generation. As per the analysis, square of the nugget diameter is almost proportional to the square of the mean current and the nugget volume increases proportional to the rate of overall heat generation. They performed simulations to predict time behavior of the temperature and voltage distribution in the weldment for various heat inputs.

Tsai et al. [19] analyzed the transient thermal responses during the process on the materials and the mechanical behavioral change during the joining. A two-dimensional finite element mesh structure was used for the analysis considering three element types:

- thermoelectric solid element to account for resistance heating and to calculate the temperature history and distribution during weld cycles
- Isoperimetric element was used to analyze stress developed from thermal strains and electrode squeezing.
- Surface element with its thickness equal to a typical oxide thickness was used to simulate the coupling effects of the thermomechanical phenomenon between electrode/workpiece and workpiece/workpiece.

It was found that when welding is done for unequal thickness sheets, the weld nugget formed is mostly in the thicker workpiece as depicted in Fig. 3.23 and

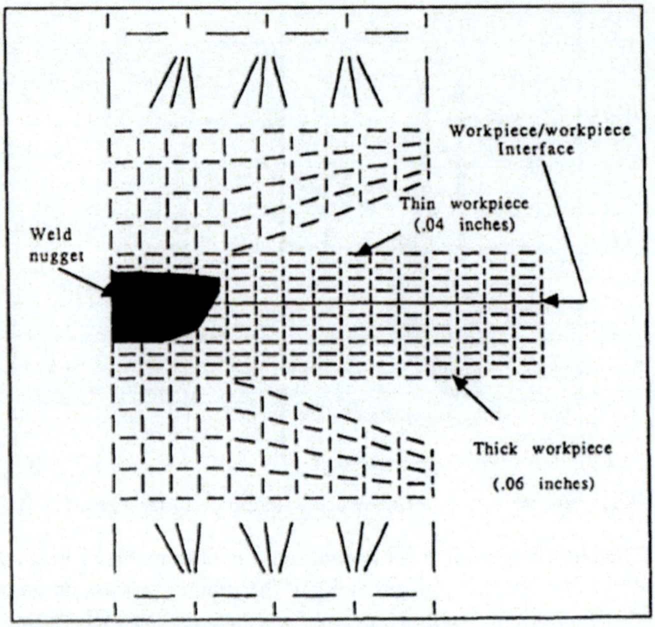

Fig. 3.23 Weld geometry for unequal thickness sheets [19]

during dissimilar materials joining, the nugget formed more in the workpiece with lower thermal conductivity or higher electrical resistivity.

The stress distribution as shown in Fig. 3.24 suggests that the maximum compressive stress obtained along the electrode/workpiece interface is near the center.

In the aforementioned publications, thermomechanical coupling and power supply parametric dependencies have not been dealt with together to get the

Fig. 3.24 Stress distribution along electrode workpiece interface [19]

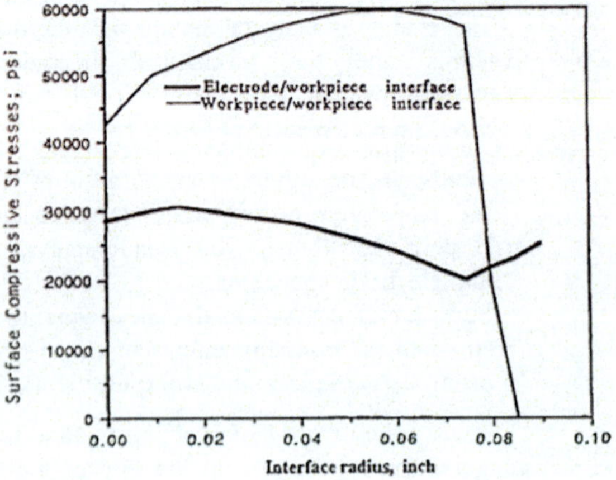

complete understanding of the interdependencies and characterization of the weld product. The mathematical models reported in the literature used for previous research works have focused exclusively on thermal or mechanical or power supply effects on the weld geometry.

3.7 Sample Design Data, Process Parameters and Design Calculations

3.7.1 Case Study for Design Calculations

Designing of a power electronics-based power source for the spot welding of 2-mm-thick 2205 duplex stainless steel sheets and 2-mm-thick AISI 316L austenitic stainless steel dissimilar material joining.

Joining dissimilar base metals is very common in the mechanical assemblies of various components in boilers, petrochemical industries, and automobile sectors. Duplex stainless steel 2205 and AISI Type 316L austenitic stainless steel are few of the mostly preferred engineering materials in industries like chemical, wastewater, marine engineering fields and desalination industries for their ease of workability and versatile features.

AISI 316L can be easily employed in caustic environments, but has high sensitivity to chloride induced stress-corrosion cracking. Duplex stainless steel (DSS) exhibits an austenite–ferrite dual-phase structure. It presents many benefits upon single-phase grades, such as increased yield strength and resistance against stress-corrosion cracking.

For automobile industries, AISI 316L and DSS 2205 are preferred for various components and assemblies. For achieving a consistent weld, appropriate power supply design is essential since weld time, current, and electrode force are the vital factors to obtain the desired quality weld.

3.7.1.1 Design Calculations—Based on Material Properties

Material selected for application in automobile industry:

2205 duplex SS sheets and AISI 316L austenitic SS
Sheet thickness (t) = 2 mm for both the sheets.

Material properties are mentioned in Table 3.1:

Power electronics-based power supply is implemented with a buck converter as shown in Fig. 3.25.

Fig. 3.25 Standard buck converter

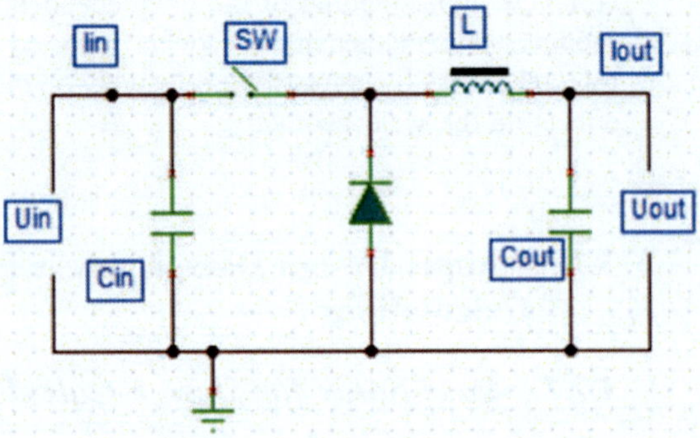

Table 3.1 Material properties of AISI 316L and DSS 2205

Material—Property	AISI 316L	DSS 2205
Specific heat (J/kg °K)	450	418
Density (gm/cm^3)	7.9	7.82
Melting temperature (°C)	1440	1465
Electrical resistivity (μΩ-cm)	74	85
Latent heat (kJ/kg)	285	285

The parameters required for the design of the buck converters are:

- Input voltage range
- Nominal output voltage
- Maximum output current required for the weld.

Below are the calculations performed based on the material selected for the RSW process described as equation (Eqs. 3.3–3.19):

$$\text{Electrode diameter} = 5 \times \sqrt{t} = 7.07\,\text{mm} \tag{3.3}$$

t—thickness of metal sheet (mm).

$$\text{Nugget diameter} = 4 \times \sqrt{t} = 5.65\,\text{mm} \tag{3.4}$$

$$\text{Area of nugget} = \frac{\pi}{4} \times d^2 = 25.06\,\text{mm}^2 \tag{3.5}$$

$$\text{Nugget Volume} = \text{area of nugget} * \text{sheet thickness} = 5 \times 10^{-8}\,\text{m}^3 \tag{3.6}$$

$$\text{Mass of the nugget} = \text{nugget volume} * \text{density} = 3.91 \times 10^{-4}\,\text{kg} \tag{3.7}$$

$$\text{Heat required to melt the nugget} = I^2Rt = m\left[C_p\left(m_\text{p} - T_\text{R}\right) + L_\text{H}\right] = 250\,\text{J} \quad (3.8)$$

m = mass of nugget, C_p = specific heat, m_p = melting point, T_R = ambient temperature, L_H=latent heat.

$$\text{Contact Resistance } R = (\rho_1 + \rho_2) \cdot \frac{1}{a} = 2.53\text{e}^{-04}\,\Omega = 0.253\,\text{m}\Omega \quad (3.9)$$

$$\text{Total Resistance} = \text{bulk} + \text{contact resistance will be approximately} = 0.75\,\text{m}\Omega \quad (3.10)$$

From the value of heat generated and the resistance, current I = 2151.65 A.

According to lobe curve generator [5], for nugget diameter of 5.65 mm:

Weld cycles = 9
At 50 Hz frequency, weld time = 0.18 s.

$$\text{heat energy } P \cdot t = v \cdot i \cdot t \quad (3.11)$$

Voltage at the weld can be found using the above relation. Voltage value is nearly = 1 V.

Input voltage is switched on and off with a pulse

$$\text{Duty Cycle of } D = (V_\text{out} + V_d)/V_\text{in} \quad (3.12)$$

Calculation of maximum switch current [20, 21]:

$$\text{Maximum Duty Cycle } D = V_\text{out}/(V_\text{in\_max} * \eta) \quad (3.13)$$

η = efficiency of the converter ≈90%.

Efficiency is added to the duty cycle calculation, because the converter also has to deliver the energy dissipated.

$$\text{Maximum Switch Current Max.}I_\text{sw} = \frac{\Delta I_l}{2} + I_{l\text{max}} \quad (3.14)$$

ΔI_l—inductor ripple current.

Values of LC filter for buck regulator:

$$L = \frac{V_{out}(V_{in} - V_{out})}{\Delta I_l \times f_s \times V_{in}} = 333\,\mu\text{H} \tag{3.15}$$

ΔI_l—Ripple current is 20–40% of output current
f_s—switching frequency 100 kHz.

Input capacitor is normally given in the data sheet. This minimum value is necessary to stabilize the input voltage due to the peak current requirement of a switching power supply.

The best practice is to use low-equivalent series resistance (ESR) ceramic capacitors.

$$\text{Output Capacitor Value } C = \frac{\Delta I_l}{8 f_s \times \Delta V_{out}} = 2.2\,\mu\text{F} \tag{3.16}$$

ΔV_{out}—Output voltage ripple.

Ratings of the Schottky Diode:

$$\text{Average Forward Current } I_f = I_{out}(\text{maximum}) * (1 - D) \tag{3.17}$$

I_{out}(maximum)—maximum output current necessary for the application.

$$\text{Power Dissipation of the Diode } P_d = I_f * V_f \tag{3.18}$$

V_f—forward voltage of the rectified diode.

$$\text{Electrode Force} = 6000(t_1 + t_2)\,\text{lbs} \tag{3.19}$$

t_1, t_2—thickness of the workpieces.

Other Observations:

Simulation techniques can be used to compare the observations with sample experimental values to arrive at optimum process parameters. The variation in the workpiece temperature and dynamic resistance has a significant effect on the weld process parameters. This variation and its effects on the process parameters can be observed in the simulation and can be further used to decide upon the required power supply characteristics.

The simulation study can be carried out to observe the following:

- Effect of dynamic resistance on the process parameters
- Variation of weld current and voltage with time for different duty cycles
- Variation of temperature with time using Foster and Cauer models
- Variation of energy losses with time
- Effect of varying workpiece temperature on the energy losses.

The observations from this study can be used to get an understanding of the RSW process and the required power supply design that can be appropriately used for various materials and weld geometries.

3.7.2 Sample Data for a Particular Application

Application—Body and panel applications in Automobile industry [11]

Material for welding: Cold Rolled Closed Annealing Sheet Metal—This is a steel grade in which cold rolling is done after hot rolling and pickling, to reduce the thickness of steel. Cold rolling makes the material hard and gives a uniform surface quality. This cold rolled steel is then annealed in a closed container where nitrogen or any other non-oxidizing gases softens it for use and also guards it against oxidation. The close annealing transforms the lamellar pearlite to spheroidal cementite and considerably develops ductility. The close annealing process also improves other mechanical properties, for example, strain hardening coefficient and planar anisotropy.

For RSW of this material with 2 mm thickness, the design specifications used for this work are mentioned:

Electrode diameter—8 mm
Nugget diameter—5.65 mm
Nugget area—$2.463e^{-5}$ m^2
Nugget volume—$5e^{-8}$ m^3
Mass—$3.95e^{-4}$ kg
Heat required for melting—273.5 kg
Only 40% of generated heat gets transferred to the interface.
Transformer Vp:Vs—440:2.8
Current requirement—3928 A at 50 Hz
Weld time—0.3 s.

3.8 Conspectus of Design Studies in Resistance Spot Welding

Involvement of myriad considerations underlines the fact that good welding is an art. Though not fiendishly complicated, mastering the process entails ages of practice and immense patience. Issues pertaining to welding have been around ever since it was invented. Factors responsible for the technical issues and the subsequent aggravations of the undesired behaviors of weld are attributed to the following:

- ignorance of weld processes
- improper weld design
- erroneous material choice and
- abysmal advised training and poor workmanship.

Weld defects are any of the unnumbered imperfections that constrict the applicability of a welded joint. Intense surveys on weld defects and their causes through case studies made us realize the necessity for bringing out this book volume that may partly address these issues with contents on counter design strategies to overcome the flaws and attain desired welds. Welding processes from different categories have been chosen for demonstrating the design aspects in this book. This chapter in particular discusses the RSW process and the associated phenomena. Besides, it presents the basic design principles and practices and weld procedure specifications along with applications and suitable materials. The next chapter takes the role of a conveyor to transfer the design aspects to TIG welding while retaining the content essence of this chapter.

References

1. Feujofack Kemda BV, Barka N, Jahazi M, Osmani D (2019) Optimization of resistance spot welding process applied to A36 mild steel and hot dipped galvanized steel based on hardness and nugget geometry. Int J Adv Manufact Technol
2. Hernandez VHB, Panda SK, Okita Y, Zhou NY (2010) A study on heat affected zone softening in resistance spot welded dual phase steel by nanoindentation. J Mater Sci 45(6): 1638–1647
3. Charde N (2012) Effect of spot welding variables on nugget size and bond strength of 304 austenitic stainless steel. Australas Weld J 57(3):39–44
4. Triyono T, Purwaningrum Y, Chamid I (2013) Critical nugget diameter of resistance spot welded stiffened thin plate structure. Mod Appl Sci 7(7):17–22
5. Asari R (2019) Resistance spot welding-weldability lobe simulation development. Sch J Appl Sci Res 2:01–05
6. Salem M (2011) Control and power supply for resistance spot welding (RSW)
7. Brown LJ, Lin J (2005) Power supply designed for small-scale resistance spot welding. American Welding Society, Dallas, Texas, 25–28 April
8. Entron resistance welding basics. https://www.entroncontrols.com/images/downloads/700081C.pdf

9. America AM (2013) Fundamentals of small parts resistance welding
10. https://www.swantec.com/technology/resistance-welding/
11. Prashanthkumar VK, Venkataram N, Mahesh NS (2014) Process parameter selection for resistance spot welding through thermal analysis of 2 mm CRCA sheets. Procedia Mater Sci 5:369–378
12. Hasanbaşoğlu A, Kaçar R (2007) Resistance spot weldability of dissimilar materials (AISI 316L–DIN EN 10130-99 steels). Mat Des 28(6):1794–1800
13. Bondarenko O, Verbytskyi I, Prokopets V, Kaloshyn O, Spitsyn D, Ryzhakova T, Kozhushko Y (2017) Modular power supply for micro resistance welding. Electr Control Commun Eng 12(1):20–26
14. Mehta ND, Haque AM Design of resistance spot welding system using integral switching cycle control technique
15. Saleem J (2012) Power electronics for resistance spot welding equipment. Doctoral dissertation, Mid Sweden University
16. Yeung KS, Thornton PH (1999) Transient thermal analysis of spot welding electrodes. Weld J-New York, 78:1-s
17. Podržaj P, Simončič S (2011) Resistance spot welding control based on fuzzy logic. Int J Adv Manufact Technol 52(9–12):959–967
18. Cho HS, Cho YJ (1989) A study of the thermal behavior in resistance spot welds. Weld J 68 (6):236s–244s
19. Tsai CL, Papritan JC, Dickinson DW, Jammal O (1992) Modeling of resistance spot weld nugget growth. Weld J (USA) 71(2):47
20. Hauke B (2011) Basic calculation of a buck converter's power stage. Texas Instruments, Application report, SLVA477B
21. Tucker J (2008) Understanding output voltage limitations of DC/DC buck converters. Analog Appl J

Chapter 4
Tungsten Inert Gas Welding and Design

Abstract TIG welding is an arc welding process that uses a non-consumable tungsten electrode for the weld and is adopted in some industries as a replacement of gas and manual metal arc welding. This is attributed to the fact that it uses inert gas shield to protect the weld pool. They are specifically preferred for joining magnesium and aluminum. TIG welding technique is of high demand for its robustness and ability to deliver quality welding of thin sheet materials or for controlled penetration for pipe welds in the industrial sectors. This chapter on TIG welding starts with the introduction to the concepts and covers the appropriate materials, applications, TIG weld machine description, and inferences from significant research reports. The last section of the chapter covers the design aspects involved in this process with the help of process parametric behaviors, sample design data, and calculations following various approaches for selection of parametric ranges.

4.1 Introduction

Tungsten inert gas welding is a fusion welding technique in which a tungsten electrode in presence of an inert gas is used to melt the materials and thus join them. The operating temperature is generally over 6000 °F. An arc generated between the electrode and the workpiece causes the latter to melt and join on solidification (Fig. 4.1). To protect the weld from the atmospheric gases, this method uses shielding gas and hence named as gas tungsten arc welding (GTAW). The decision to use or not to use the filler materials to add to the weld depends on the base metal thickness and the joint design [1].

TIG welding is predominantly used to achieve high-quality welding of mild steel or thin sections of non-ferrous metals such as copper alloys, aluminum alloys, magnesium, and stainless steel [2]. The weld is caused with a tungsten electrode of 0.5–6.5 mm diameter. Argon, helium, or a combination of the two gases in various percentages is used as shielding gas to protect the weld area from atmospheric gases shown in the schematic in Fig. 4.1. An arc is introduced between the electrode and

M. Chaturvedi and S. Arungalai Vendan, *Advanced Welding Techniques*,
https://doi.org/10.1007/978-981-33-6621-3_4

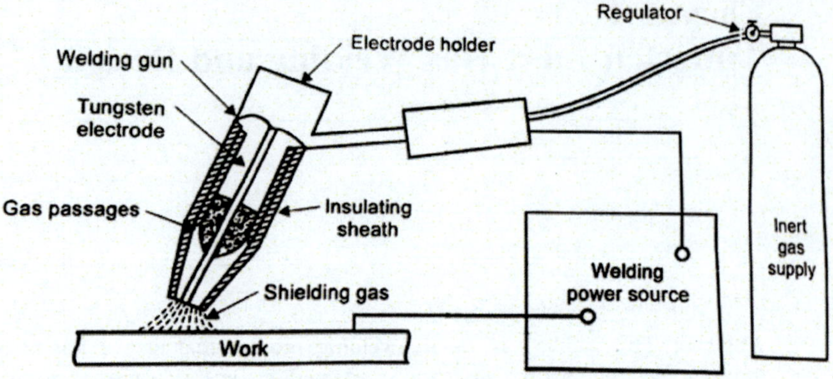

Fig. 4.1 Schematic of TIG welding [17]

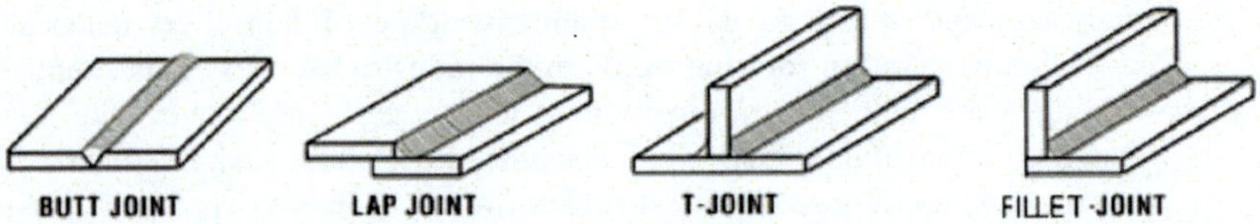

Fig. 4.2 Configuration possible with TIG welding [18, 20]

the weld plate so as to melt the material for causing the weld. The weld pool temperatures can advance up to 5432 °F

The most common joint configurations of TIG weld include the butt joint, lap joint, *T*-joint and the fillet weld as shown in Fig. 4.2.

4.2 Process Applications

The TIG welding process is generally utilized for welding of thin plates of thickness in the range of 5–6 mm with single pass weld, while using multi-pass welds for thick plates. Thus, thick plate TIG welding may result in distortions and reduction in mechanical properties, apart from high energy requirement [3].

The application of TIG welding lies mostly in the aerospace and automobile industries, especially in the welding of Al and Mg [4]. This method can be used for welding non-ferrous metals like aluminum, copper, magnesium, copper, nickel, titanium, etc.

The TIG welding process is extensively used in the following industries:

- Bicycle industry—for welding small diameter, thin-walled tubes.
- Aerospace for missiles, aircrafts, submarines, and spacecrafts—constructed in parts and for fuel pipes.
- Food processing industry because of its capability of clean and sound weld in aluminum without the use of corrosive fluxes.
- Petrochemical and Nuclear Industry.
- Automobile industry—car fenders are welded using TIG to avoid rust, safe, secure, and durable welds enable production of robust vehicles.
- Welding of exotic metals and pipeline welding.
- Industrial fixtures or metal structures for artworks and other applications.

4.3 Compatible Materials

This is the most popular method for welding Al and SS and nickel-based alloys. Due to effective control of heat input, this weld method is particularly used for thin metal parts joining. Lead, tin, or zinc alloys are not generally used with TIG welding because of extremely low melting point.

TIG welding is used for aluminum, magnesium, and for the reactive metals like titanium and zirconium. Al and Mg form refractory oxides, and Ti and Zr have tendency to become brittle on exposure to air in the weld process as they dissolve oxygen and nitrogen.

TIG welding works on higher melting point metals, and since tungsten welding operates at high temperatures, ideal metals are those that have a low melting point. Different metals have various requirements for the TIG process. Some of the requirements are mentioned here [5, 6]:

- Aluminum and magnesium: used with AC output and high-frequency setting. Cleaning of work metal required with a wire brush to remove aluminum oxide. High heat settings are required to increase welding speed.
- Copper alloys (brass, bronze, copper–nickel, copper aluminum, silicon): use DC current with electrode negative.
- Stainless steel: Filler rod with high chrome component is used, and for better gas shielding of the process, gas lenses are used with gas flow rate of 0.25–0.33 cfm
- Mild steel: Filler rods should have deoxidizers. The tungsten electrode should be 2% thoriated or 1.5% lanthanated.
- For metals having more than 4.5 mm thickness, edges need to be machined for full bead penetration.

4.4 Fundamentals of Tungsten Inert Gas Weld Process

Initiation of the welding arc requires a formation of a short circuit which is created by the scratching of the surface. Interruption of the continuous flow of current causes arc to be created. Tungsten electrode may get included in the weld due to electrode getting stuck to the surface. Lift Arc is a technique in which the short circuit is formed at a very low current level. This is used to minimize the risk of tungsten inclusion.

The most common way of initiating the TIG arc is to use high-frequency (nearly 1000 Hz) and high-amplitude (more than 1000 V) voltage sparks. Due to these high-frequency sparks, the electrode–workpiece gap gets ionized as illustrated in Fig. 4.3. Current then flows from the power source due to the electron/ion cloud formation.

Striking the arc may be done by any of the following methods:

- Touching the electrode to the work momentarily and quickly withdrawing it.
- Using an apparatus that will cause a spark to jump from the electrode to the work.
- Using an apparatus that initiates and maintains a small pilot arc, providing an ionized path for the main arc.

The electrode torch may be held manually above the workpiece or may be fixed up in an arrangement as shown in Fig. 4.3, and the placement of the electrode decides the arc length requirement. Variation of arc length by 3–4 mm can vary the voltage requirement by 5 V causing a change in the current by approximately 10 A. TIG power sources are designed to have a controlled variation of current on varying voltage, thus limiting the range of current. Consequently, small variations in the arc length do not cause observable changes in the weld indicating precise control in this process.

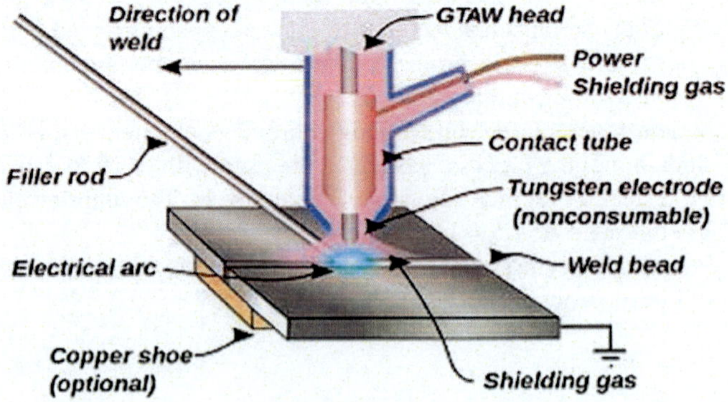

Fig. 4.3 TIG weld process [2]

The weld surface should be cleaned of contaminants before the process by the use of vapor or liquid cleaners or by mechanical means. Shielding gas (argon) also causes cleaning action at the weld surface, though helium does not have this effect.

The quality of the TIG weld gets affected by the below mentioned parameters [3]:

(a) Welding current: Selection of welding current depends upon tungsten electrode diameter, gas type, and welding polarity. High value of current in TIG welding can lead to splatter and thus result in a damaged workpiece. Lower current setting can lead to sticking of the filler wire. Optimum value of current in between these limits should be used to get an acceptable bead geometry. Bead geometry gets affected by the welding current, as the bead width decreases, height increases, and bead penetration remains constant with increase in current. In the fixed current mode, voltage gets varied to maintain a constant arc current.

(b) Welding Voltage: This is a controlling variable in manual processes because it is difficult to maintain consistent arc length. Voltage controls the length of the arc, and thus, with high initial voltage, arc initiation is easier. This also allows for a greater range of working tip distance. Although extremely high voltage can lead to unpredictable weld quality, it controls the shape of the fusion zone and weld reinforcement. Depth of penetration will be maximum at optimum arc voltage, and it directly affects the bead width. The resulting microstructure and weld quality depend upon the weld voltage.

    Influences of welding voltage on bead shape and the weld deposit composition can be listed as:

    - When there is an increase in the arc voltage, it results in a longer arc length and a correspondingly wider, flatter bead with less penetration.
    - A slight increase in the arc voltage results in the weld to bridge gaps when welding in grooves.
    - Excessive high voltage gives rise to a hat-shaped concave weld and is found to have low resistance to cracking and a tendency to undercut.
    - Lower voltages reduce the arc length, and there is an increase in penetration.
    - Excessively low voltage results in an unstable arc and a crowned bead, which has an uneven contour where it meets the plate.

(c) Inert Gases: The selection of shielding gas depends upon the weld material, cost incurred, weld speed, electrode life, arc stability, weld temperature, splatter, etc. The inert gases also affect the surface profile, finished weld penetration depth, porosity, corrosion resistance, brittleness, strength, and hardness of the weld. Argon demonstrates lesser penetration of arc compared to helium.

    Argon is preferred for extremely thin materials as it provides more control and operates at lower heat levels. Pure argon can be used for welding of structural steels, low alloyed steels, stainless steels, aluminum, copper, titanium, and magnesium.

Helium has good thermal conductivity and less electrical conductivity than argon. This, in effect, reduces the diameter of the current channel leading to current constriction. The temperature on the electrode surface is thus observed to be nearly double as compared to use of argon. It is used in TIG machines using direct current electrode negative supply for applications that involve seam welding.

Ionization energy of argon, i.e., 15.8 eV, is much lower than that of He 24.8 eV; thus, a combination of the two gases can effectively cause the ignition at larger electrode tip to weld metal distance. Helium when added with argon also increases penetration and fluidity of the weld pool. Helium argon mixtures in various ratios give differing weld results.

- 75% argon with 25% helium may be used for low alloy steels, aluminum, and copper.
- Combination of 75% helium/25% argon gives the hottest gas, but higher percentage of helium can result in arc starting issues.
- Argon with 2–5% hydrogen assists in obtaining cleaner welds without the surface oxidation since the gas acts as a reducing agent. Higher welding speeds are possible as the arc is hotter and more constricted. This mixing may cause hydrogen cracking in carbon steels and also cause weld metal porosity in aluminum alloys. This mix of gases can also be used for welding of some grades of stainless steels and nickel alloys.

(d) Welding speed: Welding speed primarily controls the bead size and penetration of weld. Slow welding speed reduces the tendency to porosity. For higher speed, heat input per unit length of weld is decreased, causing less weld reinforcement and lesser penetration of the weld. Increasing the weld speed further causes decrease in the wetting action, increases porosity, tendency of undercut, and uneven bead shapes.

(e) Material and its thickness: Properties of the material may cause a difference in the weld process design and also in the weld quality. Material properties such as thermal conductivity, coefficient of thermal expansion, reaction with atmospheric oxygen, and crack sensitivity should be taken into consideration. Its thickness helps in estimating the input heat required and required rate of cooling.

4.5 Tungsten Inert Gas Welding Machine Details

A typical TIG weld setup consists of [7, 8]:

- DC or AC/DC power source—to provide the current for welding
- Torch—to control the arc
- Control system (foot control) for amperage selection
- Shielding gas with controlled flow.

4.5.1 Power Source

The power source is a constant current power source with open-circuit voltage of 70–80 V having characteristics as shown in Fig. 4.4 [7]

Constant current power source helps avoid drawing of high short-circuit current during arc starting and during the weld, but a flat characteristic power source may damage the electrode tip or cause fusing of the electrode to the workpiece surface.

AC/DC power source may have a transformer with or without a rectifier. Current control is implemented by either a moving core within the main transformer of the power source or by using power thyristors for electronic control.

AC/DC power source is of single-phase design, while the DC source can be single-phase or three-phase design with appropriate filters. With AC power source, the polarity of the electrode and the workpiece keeps reversing in each cycle, thus distributing the heat energy on the electrode and the workpiece.

Choice of power source for TIG welding depends upon the material to be welded. Welding of aluminum and magnesium is carried out with AC power source to facilitate the removal of the strong oxide layer on these metal surfaces. AC supply enables the breaking of the oxide layer having high fusion point. For materials like stainless steel, mild steel, copper, titanium, nickel alloys, etc., DC power source can be used.

The properties and choice of AC/DC power source for TIG welding of respective materials are shown in Table 4.1 [7]

The power sources enable the following steps in the TIG weld process:

- Arc starting—initiated either by the HF spark system or by electrode scratch on the workpiece. For the weld to occur with the arc formation, high tension sparks are rapidly initiated between the electrode and the workpiece. With DC power source, once the arc is struck, the sparks cease to exist. While for AC supply, sparks continue to keep the arc alive with the positive and negative phases of power. With the HF spark system, since the electrode and the workpieces do not come in contact, thus there is no cross contamination between the two pieces.

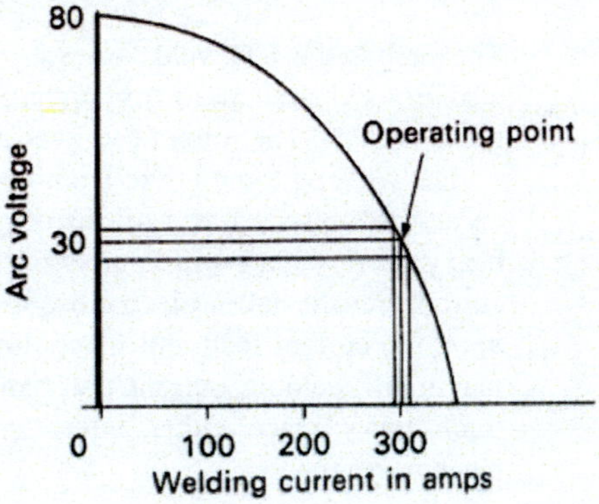

Fig. 4.4 Constant current power source [7]

Table 4.1 Properties and choice of AC/DC power source for TIG welding

Material	Type of supply —electrode polarity	Usage/reason	Weld characteristic
Steels, copper, nickel, titanium alloys	DCSP or DCEN (DC-Straight Polarity DC-Electrode Negative)	Commonly used/electrode receives 30% of welding energy	Good penetration, narrow profile
Magnesium	DCRP or DCEP (DC—reverse polarity DC-Electrode Positive)	Rare for light materials/electrode may overheat and burn	Shallow penetration, wide profile
Magnesium, aluminum	AC	Effective removal of oxides in cleaning half and weld in the penetration half	Good penetration, cleaning action
Magnesium, aluminum	Rectifier–inverter–transformer	Increased frequency reduces the transformer size	Good penetration, cleaning action
Magnesium, aluminum	AC with square wave	Better control and oxide cleaning	High penetration weld

During the short contact of the electrode with the workpiece, minimal cross-contamination occurs.

- Arc lift—Electrode in contact with the workpiece with the required control (foot or thyristor) causes a short circuit to occur, and a low current typically 5–8 A flows in the space. As the electrode is lifted off the workpiece, welding current starts to flow and the weld process continues.

4.5.2 *Torch*

The torch in the TIG weld carries the current and the shielding gas to the weld and has a control switch to allow turn on and off of the flow of welding current and shielding gas (Fig. 4.5). The torch head is coated with an electrically insulated material.

The welding torch is used to hold the tungsten electrode which provides welding current and also act as a means of guiding shielding gas in the arc zone. To avoid heavy current load on the electrode, the construction of the torch is such that the current transfer takes place close to the electrode point. Size of the torch depends upon the current load and its cooling capacity. Torches are rated according to the maximum welding current that can be used without overheating. Cooling of the torch takes place either with the shielding gas flow or with water cooling mechanisms.

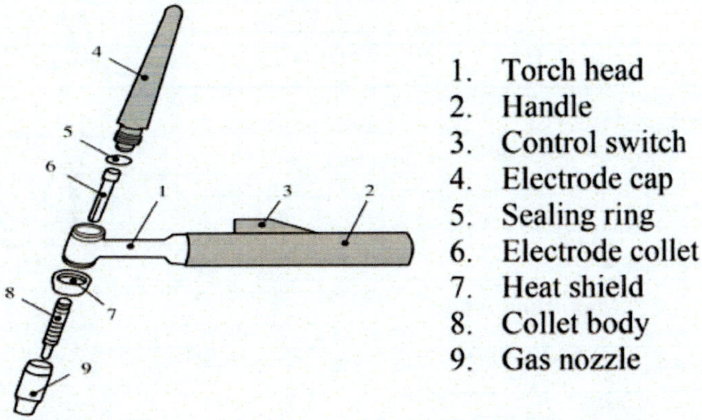

Fig. 4.5 TIG torch [19]

4.5.3 Electrodes for Welding

Tungsten electrodes are used because of the high fusion point of tungsten, which is about 3380 °C. Pure tungsten with 1–4% thorium is used to improve arc ignition for non-alloyed or low-alloyed steels. Lanthanum oxide and cerium oxide are alternative additives, which have been experimentally observed to give superior performance in arc starting and also avoid lower electrode consumption for nearly all TIG weldable materials.

Correct electrode diameter and tip angle should be selected for the range of the welding current. As a rule, smaller electrode diameter and tip angle can be used for the lower value of current; for example, the most frequently used dimensions for TIG welding electrodes are 1.6, 2.4, 3.2, and 4 mm. The grinding of the point of the tungsten electrode must be done effectively; for example, in DCEN power supply, the electrode point should be conical to obtain a concentrated arc.

In AC welding, the operating temperature of the electrode is very high, and thus, tungsten with zircon addition is used to reduce electrode erosion.

- Control arrangement:
  - For amperage selection: Welding current is controlled by any of the below mechanisms:

    Moving core in the transformers—Slow response to change due to the mechanical movement remotely controlled by a foot controller.
    Thyristor power controllers—better remote controlled based on the requirement

AC waveform balance allows the operator to adjust the amount of time for which the cleaning (positive half) and the penetration (negative cycle) take place as shown in Fig. 4.6. This is required to counter-effect the imbalance created due to self-rectification.

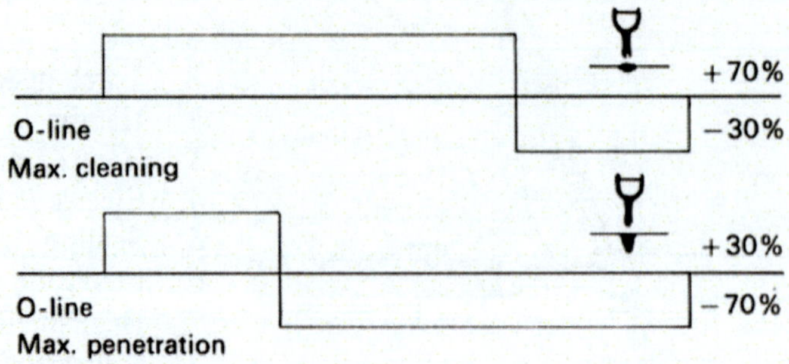

Fig. 4.6 Control of the AC waveform [8]

DC Output Pulse Control

This facility allows the welding current to be switched between preprogrammed low and high value of current. Low current of nearly 15 amps is sufficient to maintain the arc; however, it will fail to produce the necessary heat. The main current for the weld may be in the range of 50–350 A. This control can be used to provide high-quality weld and is dependent on the peak current and its time and base current and its time, and the frequency of pulses.

The pulsed current allows a series of overlapping spot welds having short cooling periods between the welds. This gives the following advantages to the weld conditions:

- Thick to thin material can be welded due to the ability to overcome the differences in heat sink
- Cylindrical or circular welds can be made without increase in the weld width
- Improved consistency in the butt welds.

Slope Control

It allows the governing of the current during the start and stop of weld as shown in Fig. 4.7. This is significant at the end of the welding as the porosity and the shrink holes can be eliminated.

4.5.4 Shielding Gases

Gas nozzles or gas lenses are used to lead the shielding gas down to the weld zone. Argon is the commonly used shielding gas for all the metals. Argon hydrogen mixtures are also used for steels and nickel-based alloys as they produce cleaner welds with deep penetrations. Helium–argon mixtures used for aluminium and copper alloys enable faster welding because of the higher arc voltage as compared pure argon.

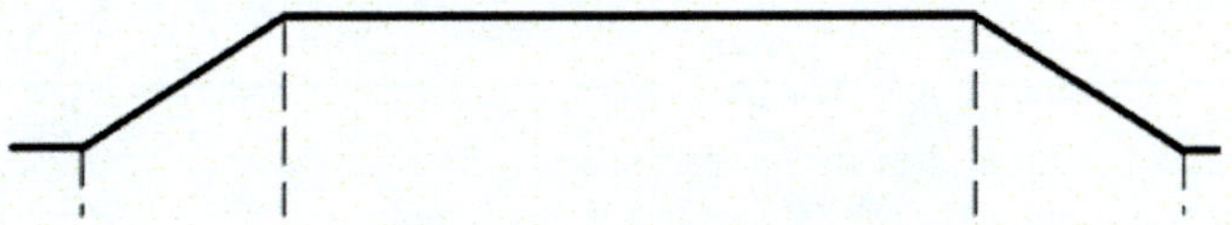

Fig. 4.7 Slope current control during start and stop [8]

Solenoid valves control the flow of the shielding gas, and the valves are electrically and PLC controlled. Argon starts flowing continuously at the rated pressure as soon as the arc is struck, and subsequently, the weld bead formation occurs. The argon continues to flow for another 5–10 s, after the arc is stopped depending on the severity of the weld. This ensures the production of the hot bead of the weld and prevention of electrode oxidation [4].

With the increase in gas flow, there is a change in bead geometry of the welded joint which dominates the weld characteristics such as weld height and weld bead [2].

4.6 Literature Survey

This section reports various observations made by researchers in their experiments for advancements in TIG welding.

Pradhan et al. [9] designed and developed an automated filler rod feeding system for TIG welding. They considered a number of mechanisms, namely rack-pinion mechanism, slider-crank mechanism, and screw–nut mechanism, and analyzed them to decide the best among the mechanisms. Screw and nut mechanism was found to give the best weld quality to suit the industry requirements. In this mechanism, linear feeding is provided by movement of screw at the end of which filler rod is connected by clamping mechanism. The screw passes through a nut which is fixed to the frame of portable moving tractor. The rotary motion of screw required for its linear displacement is given by connecting it to an external electric motor. Figure 4.8 shows the side view of the final assembly.

This proposed method has the drawback of limited distance coverage, its restriction only to linear motion and fixed angle between filler rod and the torch.

Reddy [10] investigated the experimental and the finite element method (FEM) results to analyze the weld residual stress distribution in medium thick-walled austenitic stainless steel plates SS316 plates of $0.3 \times 0.15 \times 0.01$ m. They also aimed at the TIG process parameters' selection and optimization with focus on weld sample output parameters like temperature distribution, residual stresses, and weld distortion. Variation of temperature was observed by varying welding voltage, joint efficiency, and thickness of the weld plate and found proportional variation except for weld plate thickness variation. Maximum temperature of 503 °C was observed by applying the heat flux value of 0.84×106 W/m^2, and

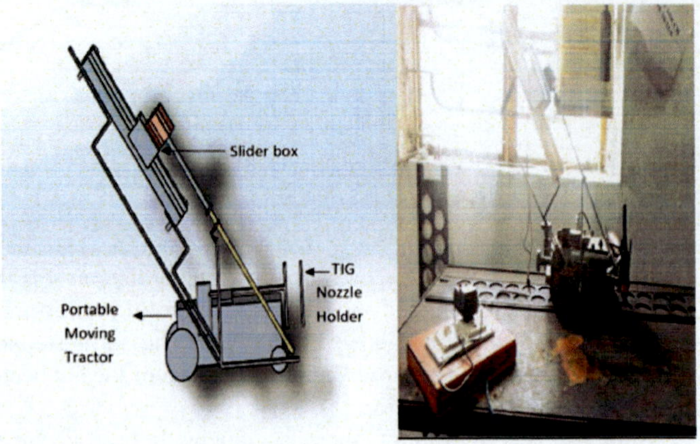

Fig. 4.8 Screw and nut mechanism assembly [9]

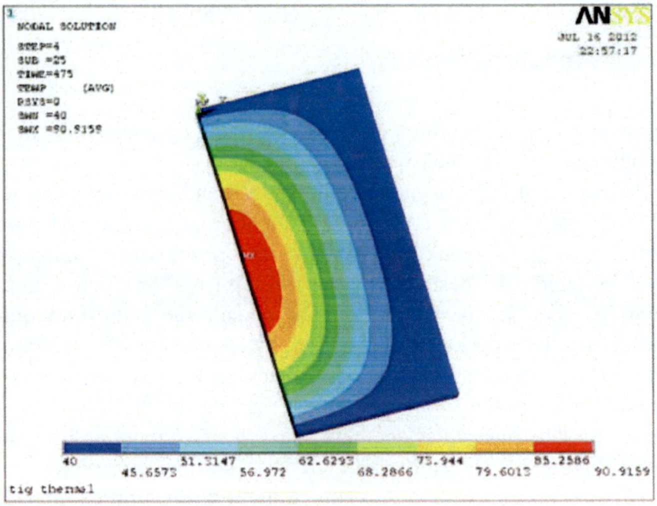

Fig. 4.9 Temperature distribution during TIG process [10]

the maximum induced stress was 136 MPa. The results shown in Figs. 4.9 and 4.10 are, respectively, for the temperature distribution and the stress distribution during the TIG process simulation.

Mohan [3] developed an automated TIG welding system to control the welding speed. Welding of commercial Al plate of thickness 3 mm was performed in two phases—single-sided weld and two-sided weld. At lower welding speeds, strength is more due to more intensity of current. For two-sided weld, tensile strength was

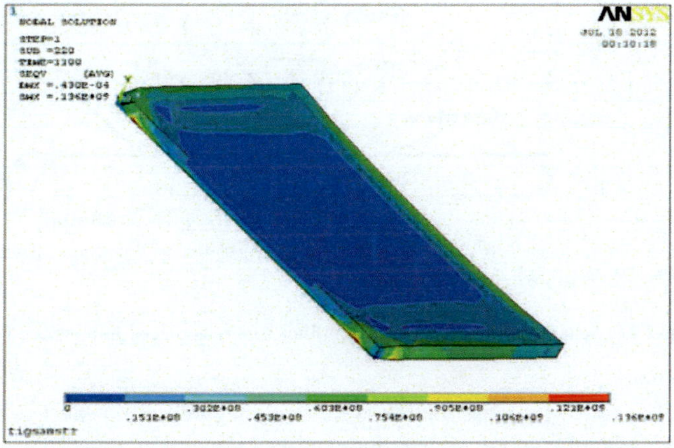

Fig. 4.10 Stress distribution during TIG process [10]

found almost equivalent to the strength of base material and with high current (180 A), welding speed has no specific effect on tensile strength of the weld joint. Hardness value of the weld zone was observed to change with the distance from weld center due to change of microstructure. At lower welding speeds, strength is more due to higher current intensity.

Lothongkum et al. [11] investigated the TIG welding of 3 mm-thick-AISI 316L stainless steel plate at different welding positions. Pure argon gas and mixture of argon with nitrogen (1–4 vol.%) were used as shielding gas with a flow rate of 8 l/min during top and back sides of welds. Effects of welding speeds and nitrogen contents in argon shielding gas on pulse currents were studied to achieve an acceptable weld bead profile with complete penetration. It was found that increasing nitrogen contents in argon gas decreases the pulse currents, and increasing welding speed will increase the pulse current.

Pujari et al. [12] worked for optimizing the weld geometry for the TIG welding of AA7075—T6 Al alloy using base plates—150 × 150 × 3.46 mm. Schematic of the process is shown in Fig. 4.11.

The Taguchi and utility concept is used as a multi-response optimization model to optimize the TIG weld process parameters on multiple performance characteristics, namely weld pool geometry parameters such as penetration, face width, and back width. The process parameters of interest are peak and base current, supply frequency, duty cycle, gas flow rate, and welding speed.

The weld groove angle must be maintained at a minimum value to reduce the loss of Zn elements from base material. Single 30° "V" groove butt joint configuration is prepared to obtain GTAW joints.

The mechanical properties of the welded joints get significantly affected by the weld pool geometry.

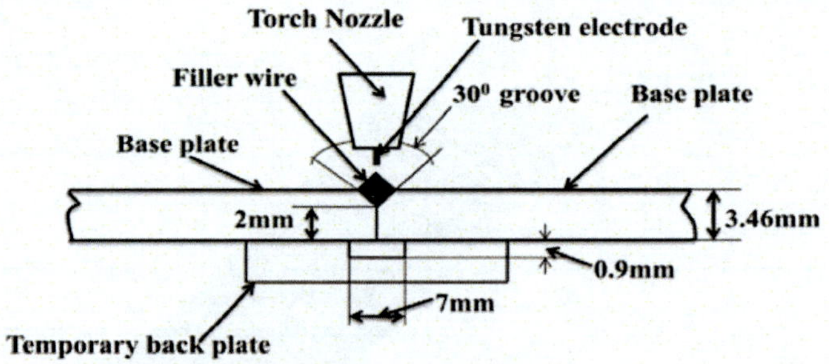

Fig. 4.11 Schematic of process showing filler materials placed in a groove [12]

The following observations were made regarding the effect of process parameters on bead contour, penetration, and weld quality:

1. For the peak current (I_p) lesser than 195 A, incomplete penetration and deficient fusion are observed. For Ip higher than 205 A, under cut, spatters, and overheating of base metal occur.
2. For base current (I_b) lesser than 93 A, arc length created is very minute, and there is no mixing of filler metal with the molten metal of base metal. For I_b higher than 103 A, arc becomes unstable, and arc length increases and starts to de-locate.
3. For the welding speed (S) less than 200 mm/min, unacceptable protrusion of the root with more undercut is recorded. For the welding speed greater than 400 mm/min, penetration decreases and weld pool becomes narrow.
4. For pulse frequency (F) less than 4 Hz, the bead contour becomes wide and poor bead quality. For F greater than 8 Hz, narrower bead contour and harsh noise is recorded with arc spatter.
5. For the pulse on time (P_{on}) of less than 40%, the heat contribution is low which is not enough to melt the base metal. If the P_{on} is greater than 60%, greater is the melting of the base and filler metal and more heating of tungsten electrode results.
6. For the gas flow rate (GF) of less than 9 l/min the arc stability, penetration and appearance of bead are poor. If GF is greater than 11 l/min, consumption and mixture of gas per kg weld deposition are extremely high. It is also observed that turbulence in the gas flow will lead to rough bead surface.

Kumar et al. [2] have studied the effects of the process parameters including the supply polarity on the tensile properties of the Al 5061 weld. They utilized the response surface methodology and a Firefly algorithm to describe and optimize TIG welding. For double V-shaped butt joint specimens of dimensions 150 mm * 150 mm * 6 mm, they observed the optimal values to be 150 A, at gas flow rate of 10 l/min and DC revere polarity (DCRP) supply connection. With these values, experimental value of tensile strength was found to be 98.89 MPa for a model predicted value of 104.12 MPa.

Prajapati [13] presented the selection of process parameters for optimal weld pool geometry of SS304 as per Taguchi design for the experiments and used genetic algorithm for optimization. They made observations regarding dependence of bead height, bead penetration, and bead width on the welding speed, current, and gas flow rate.

Observations were:

- Bead height is inversely proportional to the welding speed and increases proportionally with current and gas flow rate (Fig. 4.12).
- Bead penetration is constant with respect to current and varies proportional to the welding speed and gas flow rate (Fig. 4.13).
- Effect of process parameters on bead width is constant with change in gas flow rate that increases with welding speed and current (Fig. 4.14).

Mathematical regression equations developed from the ANOVA and normality testing for bead width (BW), bead height (BH), and bead penetration (BP) are as described in Eqs. (4.1–4.3) [13]:

$$\mathrm{BP} = 3.60 + 0.141S + 0.00033I + 0.0450L \tag{4.1}$$

$$\mathrm{BH} = -0.719 + 0.0520S + 0.00267I + 0.00333L \tag{4.2}$$

$$\mathrm{BW} = 6.71 + 0.0807S + 0.00367I - 0.00333L \tag{4.3}$$

where S—is weld speed, I—current, L—gas flow rate.

4.7 Sample Design Data, Process Parameters, and Design Calculations

For an arc welding process, it is difficult to predict the weldment characteristics by considering the effects of single process parameter. Interaction plots help by showing combined effects of input process parameters on the bead characteristics. Also, multi-response optimization is the solution to optimize responses more than one at a time. Standard procedure to find the accepted range of process parameters is by trial & error method and by visual inspection of the welds. Weld heat input is calculated with the voltage, amperage, and the travel speed and is generally in the range of 30–70 kJ/inch for various types of carbon steels and alloys. Toughness properties in the HAZ and the weld metal depend on the weld heat input. High values of the heat input beyond the estimated endurance may result in issues like slow cooling rate leading to excessive grain growth, thereby weakening the weld. On the contrary, lower heat inputs deviating from the exquisite requisite values

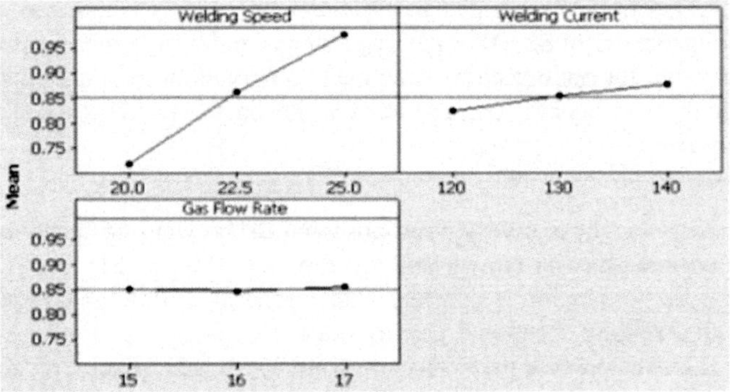

Fig. 4.12 Influence of weld speed, current, and gas flow rate on bead height [13]

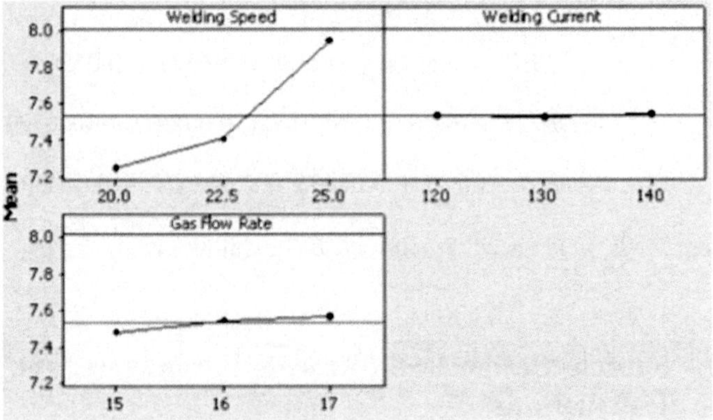

Fig. 4.13 Influence of weld speed, current and gas flow rate on bead penetration [13]

cause rapid solidification resulting in an unrefined grain structure. Eventually, the weld deposit strength is disproportionate to that of the base material. Design calculations are thus vital, and this part of the chapter presents some design methods involving vase studies for the TIG welding process.

The major TIG process operating variables are:

- Welding current, voltage, duty cycle
- Electrode composition, current carrying capacity, and shape or the tip angle—Rise in the tip angle reduces the arc voltage and also the applied power

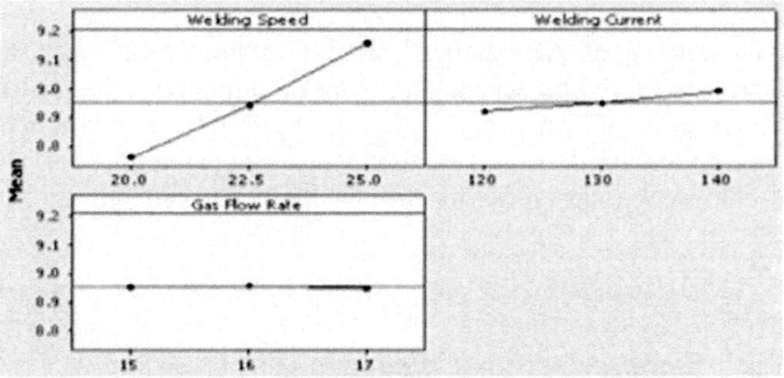

Fig. 4.14 Influence of weld speed, current and gas flow rate on bead width [13]

- Shielding gas—welding grade argon, helium, or mixtures of both—changing the nature of the shielding gas modifies the arc voltage and thus changes the applied power.
- Filler metals that are generally similar to the metal being joined and suitable for the intended service.

The previous studies have established that the depth-to-width ratio of the weld pool may get impacted with the amperage, arc length, cathode tip angle, or the shielding gas. The combined effect of all the parameters in different proportions results in different behaviors as it modifies the balance between various forces applied to the molten pool. Thus, for the optimization of the TIG process, adopting the trial and predictive approach in deciding the operable process parametric ranges may be discouraged and appropriate design process be espoused.

4.7.1 Welding Amperage Selection

Setting the output current value of the TIG machine depends on the base material and the intended weld application. Higher current input may lead to splatter and reduces the time for weld, and lower current input may lead to sticking of the filler wire. Amperage in TIG machine is controlled with the foot pedal on the machine as per the requirement. The foot pedal starts at zero amperes in the initial stage, and then, the current level can be gradually increased up to the machine's maximum current capability.

- Welding application: Application sensitivity guides the amperage in a weld setup. For example, requirements for a boiler in a nuclear power plant will be different than the requirements of a pipe weld.
- Base material: Type and thickness of the material, materials to be welded may be ferrous and non-ferrous materials having different current requirements due to individual melting point.

For welding of Al materials, current should be AC and should be able to compensate for the higher melting point of aluminum. The tendency of absorbing the high levels of heat requires the weld bead to be moved at a high speed.

The below listed are sets of values of input parameters for TIG welding of Al and mild steel, joint types, dimensions, and using different types of power supply Tables 4.2 and 4.3 [7]:

4.7.2 Sample Design Calculations

4.7.2.1 Case I

TIG machine used: ESAB TIG Torch BTE250M with MIG U5000i Welding generator
Material: AISI304L [14].

Typical values of arc current and voltage are 100 A and 16 V.

True heat input to the workpiece is given by Eq. (4.4)

$$Q_c = \eta * V * \frac{I}{v} \frac{\text{kJ}}{\text{mm}} \tag{4.4}$$

Table 4.2 Parameter values for manual TIG welding of AL using HF AC supply

Metal thickness (mm)	Joint type	Electrode diameter (mm)	Filler rod diameter (mm)	Amperage (A)	Argon gas flow rate (CFH)
1.6	Butt Lap Corner Fillet	1.6	1.6	60–85 70–90 60–85 75–100	10
3.2	Butt Lap Corner Fillet	2.4 Or 3.2	2.4	125–150 130–160 120–140 130–160	15
4.8	Butt Lap Corner Fillet	3.2 or 4.0	3.2	180–225 190–240 180–225 190–240	15
6.3	Butt Lap Corner Fillet	4.0 or 4.8	4.8	240–280 250–320 240–280 250-320	20

Table 4.3 Parameter values for manual TIG welding of MS using DC supply

Metal thickness (mm)	Joint type	Electrode diameter (mm)	Filler rod diameter (mm)	Amperage (A)	Argon gas flow rate (CFH)
1.6	Butt Lap Corner Fillet	1.6	1.6	60–70 70–90 60–70 70–90	15
3.2	Butt Lap Corner Fillet	1.6 or 2.4	2.4	80–100 90–115 80–100 90–115	15
4.8	Butt Lap Corner Fillet	2.4	3.2	115–135 140–165 115–135 140–170	20
6.3	Butt Lap Corner Fillet	3.2	4.0	160–175 170–200 160–175 175–210	20

η—efficiency, V—applied voltage, I—current and v—weld tool speed.

40% of the heat input is absorbed by the metal which causes the fusion. Remaining part of input is lost either as radiation or by convection at the anode surface.

The applied energy diffuses in the material following its temperature depending on its thermal diffusivity.

Energy required to fuse the required volume of metal is given by Eq. (4.5):

$$E_{\text{fusion}} = m * C_{\text{p}}(\Delta T) + m * L_{\text{h}} = m\left[C_{\text{p}}\Delta T + L_{\text{h}}\right) \quad (4.5)$$

m—mass of the weld metal, C_p—specific heat, ΔT—change in temperature, L_h—latent heat of the metal.

For AISI304L: melting temperature = 1724 K, density = 8gm/m^3, C_p = 0.5 J/g °C.

Pure iron boils at about 3000 K and other constituents of steel boil at lower temperature of around 2000 K. The presence of vapor due to boiling affects the interaction between the arc and the workpiece.

Accounting for the vaporization of the material, intermediate temperature may be considered.

$$T_{\text{inter}} = 2400 \text{ K}$$

Considering the weld pool as a portion of a sphere, the volume will be given by Eq. (4.6) [14]:

$$\text{Volume} = \frac{\pi}{6} d \left[3 \left(\frac{w}{2} \right)^2 + d^2 \right] \text{mm}^3 \tag{4.6}$$

For the weld dimensions as measured in sample piece:

after a time of 4 s: $d = 2.8$ mm, $w = 6.4$ mm
after a time of 20 s: $d = 3.95$ mm, $w = 7.9$ mm
Applied energy is given by Eq. (4.7):

$$E_{\text{applied}} = VIt = 16 * 100 * 4 = 6400 \text{ J} \tag{4.7}$$

Energy required for fusion is given by Eq. (4.8)

$$E_{\text{reqd}\cdot} = \rho * (\text{volume}) * C_{\text{p}} * \Delta T \tag{4.8}$$

Using the above data, volume of melt poot after 4 s volume $= 6.8 * 10^{-2}\text{cm}^3$.
On solving for energy required: $E_{\text{reqd}} = 578$ J

The applied energy from the machine and the energy required by the material can thus be determined and the process parameters be accordingly controlled. The duration for the weld can be decided on the basis of the required weld depth. The current required can be adjusted with the foot pedal on the TIG machine keeping the consideration that only 40% of the applied energy reaches the material. Also, with high current used in the process, the pressure associated with the plasma jet flow could disturb the surface tension on the weld pool causing Marangoni effect.

4.7.2.2 Case II

Material: AISI304L, 3.8 mm thick, filler metal ER 308LSi SS solid wire, diameter 0.8 mm [1, 15]
Shielding gas: argon at 12 l/min
CNC machine—to control the traveling speed and the arc length.
External wire feed machine—controls the wire feed rate.

Pulsed current TIG welding with pulse frequency of 6 and 1000 Hz.

Minimum HAZ width is the requisite characteristic of this fusion joining process. Employing pulsed current welding method helps achieve marrow HAZ. This is attributed to the fact that high frequency may cause shrinkage and shortening of the root radius of the arc.

In this technique, pulse current is supplied to create appropriate penetration depth with the fusion of the filler and the base metal. Range of base current is chosen to maintain a stable arc.

It has been observed by the previous research reports that the pulsed current improves the mechanical properties of the welded parts by significantly affecting the grain size due to breaking of the dendrite arms.

The weld pool characteristics are affected by the arc energy, its profile, and the arc pressure. The previous researches have shown dependence of the arc pressure on the square of the welding current and the pulse frequency [15].

The arc pressure, determined by the plasma jet force, affects the depth of weld penetration into the weld pool. Several other forces that act on the weld pool are gravity, surface tension, and electromagnetic forces (Fig. 4.15).

Pulsed current used in this process is shown (Fig. 4.16).

In the pulsed current scheme,
Pulse current—$I_p(A)$
Base current $I_b = 5 - 95\%\,\text{of}\,I_p(A)$
Mean current given in Eq. (4.9)

$$I_m = \frac{I_p t_p + I_b t_b}{t_p + t_b} \tag{4.9}$$

Peak current duration t_p (ms)
Base current duration t_b(ms)
Cycle time given in Eq. (4.10)

$$T = t_p + t_b \tag{4.10}$$

Pulse frequency as given in Eq. (4.11)

$$F = \frac{1}{T} \tag{4.11}$$

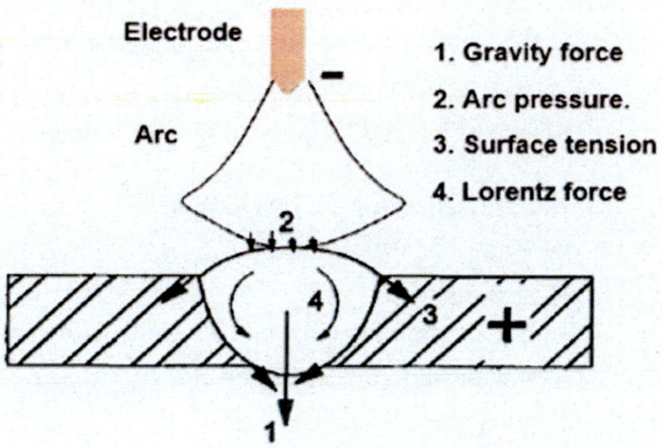

Fig. 4.15 Forces influencing the weld pool [15]

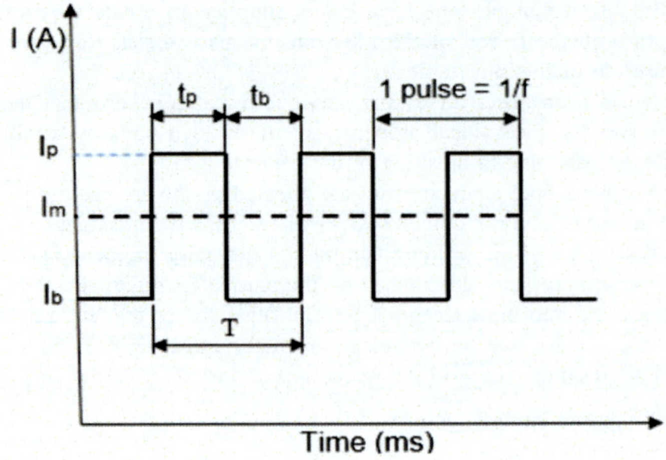

Fig. 4.16 Pulsed current time waveform [15]

Range of process parameters can be decided based on preliminary trial data.
Sample values: weld current—175 A
Peak time = 50%, arc length = 4 mm,
Base current = 90 A wire feed angle = 40°.

Either by the use of preliminary data or with optimization techniques, the optimum values of arc travel speed, wire feed speed, and the pulse frequency can be determined.

$$\text{Arc travel Speed } S_{\text{arc}} = \frac{2\text{ mm}}{\text{s}}; \text{Wire Feed speed} = \frac{33\text{ mm}}{\text{s}}, \text{frequency} = 1000\text{Hz}$$

Heat input per unit distance to the weld pool determined by Eq. (4.12)

$$H_{\text{i}} = \frac{\eta * I * V_{\text{av}}}{S_{\text{arc}}} \left(\frac{\text{J}}{\text{mm}}\right) \tag{4.12}$$

η—arc efficiency, V_{av}—average arc voltage

Arc efficiency for TIG weld = 83%
Average arc voltage = 15 V

Heat input:
For continuous current TIG weld with $I = 175$ A, $H_{\text{i}} = 1089.5$ J/mm
For pulsed current TIG weld, $I_{\text{m}} = 132.5$ A, $H_{\text{i}} = 824.81 \frac{\text{J}}{\text{mm}}$

$$\text{Arc Force} \quad P_{\text{arc}} = \frac{\mu_0 I^2}{4\pi^2 R^2} \tag{4.13}$$

μ_0—space permeability, I—weld current, R—arc radius [15].

For pulsed current TIG weld, $I = I_m$ – mean current and Frequency $f = \frac{1}{t_p + t_b}$ Hz.

Using the above equations, equation for arc pressure changes to the form given in Eq. (4.14):

$$P_{\text{arc}} = \frac{\mu_0}{4\pi^2 R^2} * f^2 I_b^2 (Ct_p + t_b)^2 \tag{4.14}$$

I_b—base current, C—assume to be 2

P_{arc} = 1265.625 W for pulse frequency f = 1000 Hz.

The arc pressure increases with increasing frequency. The pulsed arc pressure produces oscillation causing breakage of dendrite arms resulting in smaller grain size and refined structure. This affects the tensile strength of the welded parts. Thus, estimating the range of required arc pressure from the tensile strength requirement of the application or from preliminary trials, the process parameters for the pulsed TIG welding can be decided.

4.7.3 Sample Data for Various Materials and Applications

4.7.3.1 Set I

Material—commercial Al 3 mm thickness, 120 mm * 50 mm [3].
Current type—AC since it concentrates the heat in the welding area.
Electrodes—zirconated tungsten diameter—3.5 mm, tip diameter—2.26 mm

Experiment trials help to arrive at appropriate parameter range for good quality weld
Welding current (100–140) A, voltage 50 V
Speed (3.5–4) mm/s
Distance of tip from weld center is 3 mm
Gas flow rate (8–10) l/min.

4.7.3.2 Set II

Material dimensions—mild steel plates: 180 mm * 65 mm * 8 mm [16]:
Power source—direct current straight polarity (DCSP) with arc image magnifying system. The experiment trials that yielded defect-free weldments were conducted to select the range of the inputs process parameters.

Linear variable differential transformer and arc magnifying system are used to set the arc length.

Current—55–95 A
Speed—15–45 mm/min

Current capacity depends on the electrode sizes. Observations made were as in Table 4.4.

4.7.3.3 Set III

Material—SS316 plates of 300 * 150 * 10 mm dimensions [10]
Supply voltage—30–450 V, three phase, 50 Hz
Machine rating—30kVA
Input current—40–80A
Welding voltage—DCEN–15 V
Weld speed—2.5 mm/s
Electrode—2% thoriated tungsten
Shielding gas—argon

For a power supply of 15 V, 80 A, the heat flux (q) calculation is shown Eq. (4.15):

$$q = \frac{Q}{A} : \text{where} Q = V * I * \eta = 15 * 80 * 0.7 = 840\ \text{W} \tag{4.15}$$

For a cross-sectional area $A = 4.5 * 300 * 10^{-6}\ \text{m}^2$, $q = 0.62 * 10^6\ \text{W/m}^2$.

Table 4.4 Observation table for input parameters and the resulting bead width

SI. No	Current (A)	Arc length (mm)	Speed (mm/min)	Experimental bead width (mm)
1	60	1.5	15	5.17
2	65	2.3	23	6.21
3	95	1.8	27	6.41
4	95	3.2	35	7.06
5	110	2.3	42	6.33

As the efficiency of heat generation varies, the heat flux will vary which will cause variation in the temperature distribution in the weld region. Thus, selection of the process parameters ranges can be performed to achieve the required temperature for the welding.

4.8 Conspectus of Design in Tungsten Inert Gas Welding

An attempt is made in this chapter to evaluate the design aspects of the TIG welding process in terms of welding parameters and input energy consumed per weld. This may help engineers to forecast cost per welding, productivity, and quality assurance. In parametric evaluation, the actual time for each welding, heat input, arc parameters, and associated strength parameters are computed. This evaluation facilitates a comparison as to which welding process is more suitable for welding any specific material for specific application. This type of information is useful in design and production planning when engineers are often faced with difficulty in choices. This information contained here will help in establishing the suitability of TIG welding for various applications. In addition, it will help engineers to avoid making what could be costly mistakes or to overcome problems when they occur in production. Findings on joint designs discussed in Chaps. 3 and 4 provide understanding and broad guidelines in designing weldments involved in conventional RSW and TIG processes for a particular application. Optimization of weld parameters to obtain high strength and enhanced quality of the weldment for a particular material and dimension is also elaborated for RSW and TIG. Some of these design guidelines and procedures are invariably generic to all categories of welding processes. Hence, the basics on these are essential for design engineers who may partially adopt them for advanced laser joining process to be discussed in the following chapter in addition to the other relatively newer approaches for optimizing the weld parameter. Chapter 5 addresses the design aspects involved in laser beam welding which is chosen from the high-energy beam category. The approach and treatment would remain the same, while the parameters would vary based on the process phenomena.

References

1. Quigley MBC, Richards PH, Swift-Hook DT, Gick AEF (1973) Heat flow to the workpiece from a TIG welding arc. J Phys D Appl Phys 6(18):2250
2. Kumar R et al (2017) Experimental investigation and optimization of TIG welding parameters on Aluminum 6061 alloy using firefly algorithm. IOP Conf Ser Mater Sci Eng 225:012153 https://doi.org/10.1088/1757-899x/225/1/012153
3. Mohan P (2014) Study the effects of welding parameters on TIG welding of aluminium plate (Doctoral dissertation)
4. Ravinder Reddy P. Simulations of TIG welding process (weldingsimulations_modified)

5. TVM@2017, 2019 Laser beam welding—equipment, principle, working with advantages and disadvantages. Viewed on 12 Aug 2020. https://www.theweldingmaster.com/laser-beam-welding/
6. Jeff G (2019) TIG welding (GTAW) process & how it works. Viewed 20/9/2020. https://weldguru.com/tig-welding/
7. https://www.weldability-sif.com/media/docs/Intro_TIG_Welding.pdfTIGWelding.pdf
8. http://www.weldingcourseinindia.in/WELDER_SKILL_DEVELOPMENT_COURSE_IN_CHENNAI_WELDER_SKILL_DEVELOPMENT_TRAINING_COURSES_IN_CHENNAI_WELDER_SKILL_DEVELOPMENT_TRAINING_AND_CERTIFICATION_IN_CHENNAI.html
9. Pradhan RP, Shiva Das PG (2015) Design and development of automated filler rod feeding system for TIG welding (Doctoral dissertation)
10. Reddy RP (2014) Simulation of TIG welding process
11. Lothongkum G, Viyanit E, Bhandhubanyong P (2001) Study on the effects of pulsed TIG welding parameters on delta-ferrite content, shape factor and bead quality in orbital welding of AISI 316L stainless steel plate. J Mater Process Technol 110(2):233–238
12. Pujari KS, Patil DV, Mewundi G (2018) Selection of GTAW process parameter and optimizing the weld pool geometry for AA 7075-T6 Aluminium alloy. Mat Today Proc 5 (11):25045–25055
13. Prajapati AH Experimental investigation of process parameters on weld bead geometry for SS-304 using tig welding
14. Stadler M, Masquère M, Freton P, Gonzalez JJ (2017) Experimental characterisation of the weld pool expansion in a tungsten inert gas configuration. Sci Technol Weld Join 22(4):319–326
15. Ugla AA (2018) Enhancement of weld quality of AISI 304L austenitic stainless steel using a direct current pulsed TIG arc. IOP Conf Ser Mater Sci Eng 433:012075
16. Narang HK, Mahapatra MM (2014) Statistical analysis of TIG arc weldment characteristics. Int Proc Econ Dev Res 75:73
17. http://gray.ilari.benkeme.mohammedshrine.org/gas-arc-welder-wiring-diagram.html
18. Ben Zandstra (2017) PE2BZ datasheet archive. Viewed on 29/9/20. https://pe2bz.philpem.me.uk/Lights/-%20Laser/Info-902-LaserCourse/c04-04/mod04_04.htm
19. http://coun.rosz.lopla.tixat.eumqu.hicag.momece.tivexi.tixat.mohammedshrine.org/tig-welding-torch-diagram.html
20. Advantage Fabricated Metals (2009) Advantage fabricated metals. Viewed on 20/8/20. http://www.advantagefabricatedmetals.com/tig-welding.html

Chapter 5
Laser Beam Welding and Design

Abstract Laser welding is a fusion welding process and uses a laser beam of a determined spot size to cause melting of the workpieces and the filler material to form the weld. Laser welding, when it originated had taken over as an advantageous technique over the traditional arc welding methods in high-volume manufacturing industries. It offers the key advantages of better weld strength, restricted heat affected zone, applicable to a larger variety of metals, high precision, and minimal deformation. This chapter on laser welding starts with an introduction to the concepts, requirements, and working of laser welding and covers its applications, compatible materials, and summary of significant research advancements. The chapter concludes with the focus on design perspectives of laser welding with the description on critical process parameters and their standard values based on experiment trials. The chapter also has case studies explaining the design sequence based on HAZ geometry, thermal gradient curves, optimization techniques, and sample calculations for these approaches.

5.1 Introduction

Laser beam welding (LBW) is an advanced joining technique having diverse applications from aerospace industry to fine jewelry making.

The metal components to be welded get heated up to their melting point forming a weld pool when a tiny spot along the weld line is subjected to a highly concentrated laser beam. Metals are melted with the absorbed light that causes excitation of electrons followed by bond breaking within the atoms. Joining of the two metals is initiated as the weld pool loses its temperature and solidifies with the progression of the beam on the joint [6]. Thus, the melting of the two materials at their seams forms a joint (Fig. 5.1).

High beam laser welding of materials is governed by factors like the wavelength of the laser, capacity of the material to absorb laser energy, material thickness, molecular structure, and material additives.

M. Chaturvedi and S. Arungalai Vendan, *Advanced Welding Techniques*,
https://doi.org/10.1007/978-981-33-6621-3_5

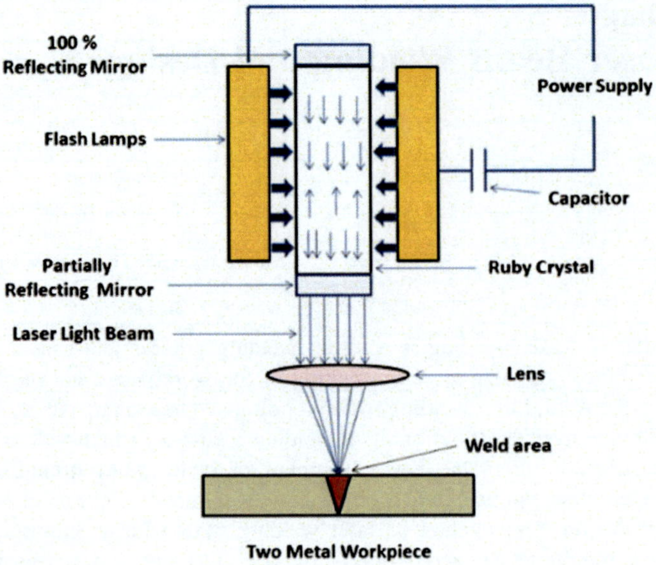

Fig. 5.1 Laser welding schematic [22]

Laser welding can be used to weld a variety of materials including carbon steels, stainless steels, titanium, aluminum-nickel alloys, and plastics. High-volume production, best quality weld, and low distortion are some of the features which establish laser beam welding method as a universal joining process in industries.

Typically, the heat inputs are lower than the arc welding processes. Laser welding involving deep penetration is chosen when welding components are expected to be free from thermal distortion or where several layers of materials have to be welded simultaneously.

LBW involving fast processing speeds typically of the order of meters per minute for sheets/plates facilitates higher productivity. A single pass of laser beam is needed for obtaining narrow, deep penetration weld between square-edged parts in thicker materials.

Laser welding has several advantages over other welding techniques:

- No tool wear
- Precision targeting
- No electrode is used
- High-quality welds
- Negligible addition in the weight at the joint making it a strong and lightweight
- Welding process can be easily automated using a CAD/CAM setup.

Conduction weld and keyhole weld are the two operating modes of laser welding. The power density of the beam decides the mode of welding [23, 24] for a particular material based upon the weld depth achievable as illustrated in Fig. 5.2.

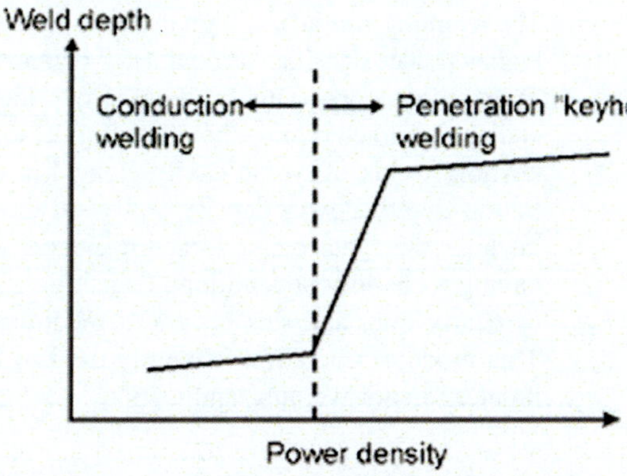

Fig. 5.2 Relationship between power density and weld depth [24]

- Conduction limited welding—In this method, the metal surface is heated beyond its melting point. Heat is controlled to prevent the metal from vaporization. The weld has lesser strength since the beam does not penetrate into the material, only surface absorption takes place as shown in the schematic of Fig. 5.3. The final weld has smooth surface and exhibits a high width to depth ratio. This mode of welding requires low power, i.e., <500 W and a power density of 0.5 MW/cm^2 [23].
- Keyhole mechanism—In this mode, the metal is heated to vaporization temperature and the laser beam penetrates into the materials forming a "keyhole'-shaped cavity. This cavity is filled with expanding metal vapor which prevents the molten metal to fall into the cavity. Some molten material still is observed to flow around the keyhole cavity under the Marangoni effect and thus cools and solidifies to form the weld. Chevron pattern is noticed at the weld cap pointing toward the start point of the weld. The keyhole-shaped cavity enhances the penetration of laser beam into the workpiece. The keyhole extends along the interface to be welded or the joint is moved with respect to the laser beam to achieve deep penetration weld.

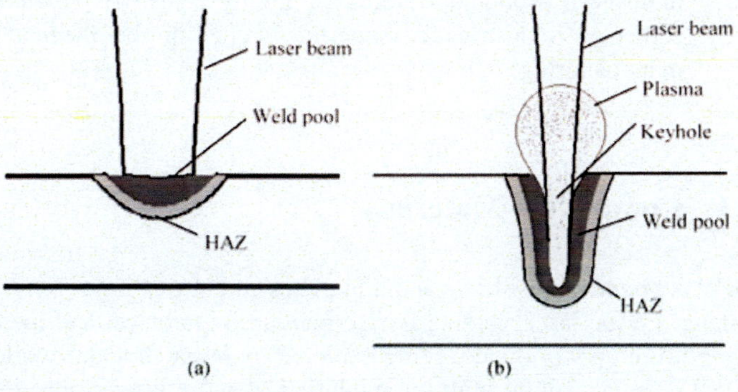

Fig. 5.3 Schematic of conduction limited and keyhole welding [23]

The resulting weld has a high depth-to-width ratio. This mode of welding requires higher power densities around 1.5 MW/cm^2.

- Transition mode—This mode employs intermediate range of power densities, and the characteristics show similitude to the ones observed conduction and keyhole welds. Keyhole cavity created in the materials results in greater penetration depth. Power density and weld time determine the keyhole depth. The keyhole that gets created cannot extend very deep into the metal plate, thus having a shallow penetration. Transition mode welds have depth-to-width ratio of nearly one, and this occurs at medium power density, around 1 MW/cm^2. This mode of operation is mainly used in Nd:YAG or fiber lasers for low heat input and spot welding applications.

5.2 Process Applications

LBW shows greater potential to be deployed in automotive industries for joining roof components, doors, or filter assemblies. The following features make LBW an effective alternate process:

- Provides the provision of single side accessibility
- Reveals an increased torsional stiffness
- Appropriate for thin-walled assemblies due to small diameter of laser beam
- Precise for compact components
- Allows for design flexibility
    - LBW is deployed in automotive industries for manufacturing components, viz. engine parts, transmission parts, alternators, solenoids, fuel injectors, fuel filters, air-conditioning equipment, air bags, and stainless steel bushes for sensors [3].
    - LBW is also implemented in the jewelry and medical industries for small components welding. Besides, it finds applications in the electronics industry, membrane welding of highly sensitive pressure sensors in the aerospace industries or superfine welds in the medical instrumentation manufacturing.

5.3 Compatible Materials

LBW is preferred for sheets and thin plates and also at times for dissimilar material joining. Plastic laser welding is also gaining importance, and intensive research is being taken up worldwide, being explored by research teams worldwide. Adoption of LBW is expanding with the evolution of more laser compatible materials and additives.

Weldability analysis helps in appropriate material selection considering the weld geometry and design, tolerances to deformations, and the application-specific requirements like strength, electrical contact, or sealing [1].

Thorough understanding of welding metallurgy is important when considering two materials to be joined owing to the possibility of formation of intermetallic regions that may lead to brittleness, thus necessitating the tensile properties testing. It is also essential to estimate the minimum heat input and laser time, respectively, to achieve effective welds. The two parts to be joined are called the transmitting part and the absorbing part.

For industries like batteries and medical applications, dissimilar material welding is required that is considered to take the properties of both materials for improved characteristics in the application and ensure weldability.

In laser welding, microweld and penetration weld may be suitable for different types of materials. Selection guideline for materials has been proposed as part of earlier research [2] and is included here (Table 5.1).

For plastic laser welding [1], the absorbing parts should be dark-colored or in shades of gray and beige. These dark-colored parts of high transmission rate materials result in faster weld production. Two optically clear parts can also be laser welded by addition of an absorptive interface.

Any thermoplastic can be welded to itself. For the melting phase dissolution, there should be an overlap range of 50 °C in the melting point when dissimilar thermoplastics have to be welded.

Some of the plastic materials that have been welded successfully include [1] polycarbonate (PC), acrylonitrile butadiene styrene (ABS), low-density polyethylene (LDPE), high-density polyethylene (HDPE), polypropylene (PP), and polymethyl methacrylate (PMMA).

Table 5.1 Materials and their properties for the weld process

Material	Weld properties
304 and 304 L series stainless steel, Al alloys, titanium	Preferred for microwelds, Al alloys should be tested for crack sensitivity
Carbon steel	Preferred for penetration welding, carbon content range different for continuous weld (up to 0.2%) and pulsed weld (less than 0.12%)
Copper	High-energy levels required to overcome surface reflectivity
Nickel alloys, nittinol, phosphor bronze	Acceptable welds, nittinol welds may be brittle
Al-cold rolled steel, Al-copper, steel-nittinol, steel-titanium	Brittle intermetallics are created, Cr/Ni ratio in weld metal made higher by having offset into steel during welding
Steel-Inconel	Sensitive to cracking
Steel-copper, copper-bronze, titanium-Al	Acceptable welds

5.4 Fundamentals of Laser Beam Weld Process

The process of laser weld involves directing a focused beam of photons to the surface of the weld material. The laser beam then gets reflected, and some part also gets transmitted or absorbed by the receiving material based on its properties. These effects can be observed in all materials in varying proportions [1].

Quality of the laser beam and the characteristics of the corresponding optics help determine the weld cross section [4].

A high-quality, or bright, laser beam provides a small, intense focus spot that facilitates deep and narrow weld for a given laser power and processing speed. Deep and narrow welds are obtained with short focal lengths. Wide weld joints are preferred for thin parts welding as the penetration is easier.

For transmitting parts, when the material transmission rate is 20% or higher, the desirable welds are obtained. Low transmission rate may result in higher process sensitivity [4].

Laser weld machines come with a shielding gas attachment, for its proper functioning at normal atmosphere. Humidity in the air can cause the production of hydrogen during the welding, and this diffusion of hydrogen into the metal leads to weak weld joints. Nitrogen may mix with the molten metal and cause the formation of voids or holes within the weld leading to weld failures.

> Laser welding is usually accomplished using high power densities of the order of 1 MW/cm^2. The spot size of the laser can vary between 0.2 and 13 mm. Power input and the focal point location control the depth of penetration. To maximize the penetration, the point of focus of the laser beam should be below the surface of the workpiece [31].

The main parts of the laser beam equipment are briefed in the next section.

5.5 Laser Welding Machine Details

A laser machine is made up of these elements arranged in blocks as shown in Fig. 5.4 [31].

- Laser medium comprises laser element doped on a host material for solid-state lasers.
- Pump source—provides the activation energy to the laser medium thus generating the laser beam.
- Laser resonator—consists of two reflectors—a rear 100% reflector and a front partial reflector and controls generation of laser in the medium.

The most common lasers used in a LB machine are briefed below. Laser beam is formed when on excitation, the laser medium starts to emit photons [31].

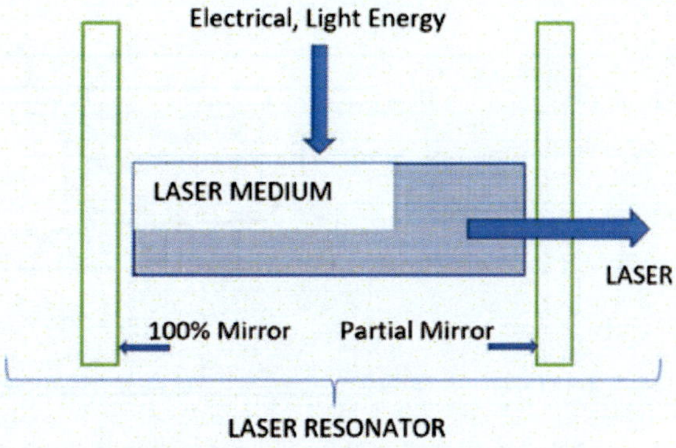

Fig. 5.4 Block arrangement for laser machine

- Gas Laser—An electric current is discharged through a gas inside the laser medium to produce laser light. Mixed gases of helium, nitrogen, and carbon dioxide are used as the medium.
  The laser medium, i.e., the gas mixture is given excitation energy with high-voltage, low-current power sources as in the arrangement as shown in Fig. 5.5.
  CO_2 gas laser beam has a wavelength of 10.6 μm which is the deep infrared region, and, in this arrangement, continuous and pulsed mode laser beam can be used to cause the heat energy transfer. A rigid lens and mirror delivery system is used for energy reflectors.
  Power outputs for gas lasers can be in the range of 25 kW which is much higher than solid-state lasers. It produces a laser light beam in the infrared region of the spectrum at 1.15 μm. These are used in applications that require laser light with very high beam quality and long coherence lengths [5].
- Solid-state laser uses glass or crystalline materials including synthetic ruby and Nd:YAG. Neodymium is a laser material which produces high levels of peak power compared to any other doping element [5, 22].
  Nd:YAG lasers can be operated in pulsed and continuous mode. Figure 5.6 shows the arrangement for Nd:YAG lasers.
  Typical power output for Nd:YAG laser is between 0.04 and 6,000 W. The beam quality is inferior to gas lasers [5].

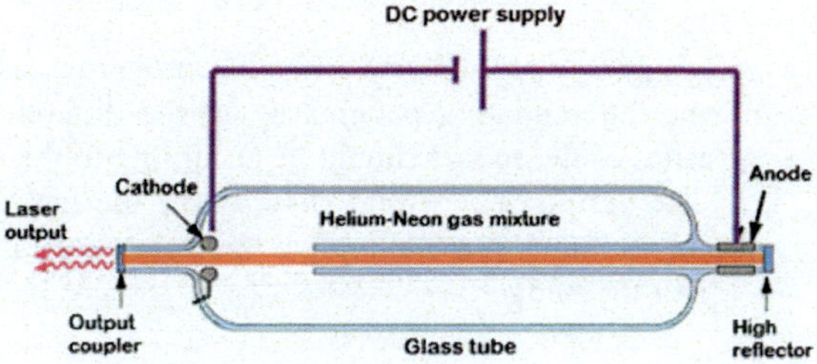

Fig. 5.5 Gas laser schematic [5]

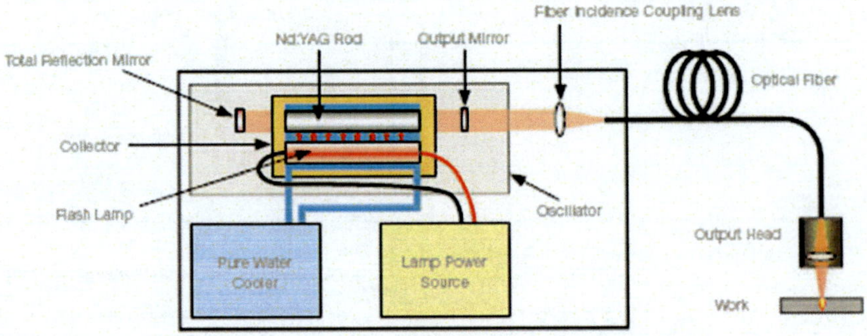

Fig. 5.6 Arrangement of Nd:YAG laser [30]

- Fiber Laser—The main medium is the optical fiber itself. The generated laser beam is emitted from optical fibers doped with an element, either Ytterbium (Yb) or Erbium (Er) depending on the desired wavelength. Doping with Ytterbium produces a wavelength between 1,030 and 1,100 nm. They are largely used for robotic industrial welding owing to the high power capability of 50 kW.
- A diode laser is a semiconductor device that directly converts electrical energy into laser light. High power diode lasers emit light in the near-infrared spectrum, 800–940 nm. A number of emitters can be fabricated on a single semiconductor substrate with a total output of nearly 100 W ensuring greater efficiency than single diode. The voltage difference across the diode stack enables laser operation with the energized charge carriers.

  Diode laser medium can be made up of a variety of materials, the most common being gallium arsenide.

  Key advantages of diode lasers are the efficiency, ability to endure, long operating hours, and compact size. A high-power diode laser is approximately 40% efficient in converting electrical energy into light and can provide warranty of 30,000 operating hours. In comparison, a lamp-pumped Nd:YAG laser requires a lamp change every 1,000 operating hours.

Diode lasers offer a distinct advantage for welding aluminum. The wavelength of diode lasers is closer to the absorption peak of aluminum in comparison to fiber lasers. Diode lasers can even be used for microwelding thin stainless steel foils, thermoplastics and also conductive ribbons in Photo Voltaic industry.

Quality of the beam and other properties like brightness and wavelength will be the differentiating parameters for the different laser sources. The build characteristics of the source should be according to the requirements of the weld application. The design of the weld should take the material properties and the build requirements into consideration like the weld penetration, joint geometry, fit up tolerances, and the budget [14].

Weld beam quality and brightness are the controlling factors for penetration of depth and the weld speeds and also have inverse relation with the weld stability and tolerance [14].

- Less than 0.01 in weld penetration—The pulsed Nd:YAG laser can be used in this category of applications which require high control and also can be used for various applications with differing requirements of weld geometries and materials.
  Fiber laser can be used to focus on to spot sizes under 25 microns and thus can be used for optimal power densities in the form of continuous weld.
- Between 0.01 and 0.03 in weld penetration—The pulsed Nd:YAG is generally used for spot welds and the fiber laser is employed for butt and filet welds. The fiber laser requires nearly 500 W power and can result in a spot diameter of 0.01 in. Pulsed Nd:YAG laser can cause equal penetration even at differing speeds while utilizing 500 W or 25 W power. The penetration depth is determined by the peak power, and weld speed for seam welds is determined by the average power.
- Weld penetration larger than 0.03 in—This is the general weld dimension in which all the forms of laser can be employed. For penetration in the range of 0.05 in, pulsed Nd:YAG is used. Other forms of lasers are used for higher depth builds up to and beyond 0.5 in. Pulsed Nd:YAG welds are used for pressure sensors involving seam welding.

- Laser beam delivery

Normally, the laser beam is supplied to the laser welding machine by the use of optical fibers. There are single fiber welding machines and multiple fiber welding machines. The multiple fiber welding machines have a laser connected to each fiber.

To concentrate the beam to a point, a collimator lens cascaded with a focusing lens is often used.

LBW machines can be classified into the following types [31]:

- Traditional: Implemented as a robot welding where the laser output follows the required seam.
- Modern: Welding is implemented remotely with a laser scanner, giving the advantages of higher speed and precision in the weld process.

5.6 Literature Survey

Input parameters of LBW are power input, speed, length of focus, spot or fiber diameter, and shielding gas flow. Each input factor contributes to the weld quality and strength for a given material These input parameters have significant impact on the build geometry, microstructure, and tensile properties of the weld, which necessitates a thorough analysis [6].

A brief summary of work done by various researchers to observe the effect of input parameters and the methodology used on the output responses is presented in this section.

Adisa et al. [7] experimented to determine the optimum value of the process parameters for the laser welding of AA7020 alloys hot rolled at a temperature of 470 °C to a thickness of 1 mm. The samples were welded with Al-5Ti-B and Al-5 Mg as filler metals. Al-5Ti-B improves the microstructure and ultimate tensile strength of AA7020 aluminum alloy.

The samples of AA7020 were butt welded with filler metals using a Nd:YAG laser. The optimum welding speed and laser peak power value were found to be 1 mm/s and 0.91 kW.

Suder et al. [8] investigated the impact of power density, interaction time, and specific point energy on the weld properties and inferred that the weld width is controlled by the interaction time, while the penetration depth is impacted by the power density and specific point energy. For a laser beam with a uniform intensity distribution and a constant interaction time across the laser spot, the specific point energy (E_{sp}) is given by:

$$E_{sp} = q_p.\tau_i.A_s = P.\tau_i \text{ Joules}$$

q_p—power density, τ_i—interaction time, A_s—surface area, and P—laser output power [8].

Analysis of the characteristics shown in Fig. 5.7 allows distinguishing various laser processes according to the power density and the interaction time, the product of which is the energy density. The area of interest being the keyhole welding, it can be read from the processing map that the laser interaction time is in the range of 10^{-6}–10^{2} s while the power density range being 10^5–10^6 W/m^2 and the operating temperature is nearly 2,750 °C.

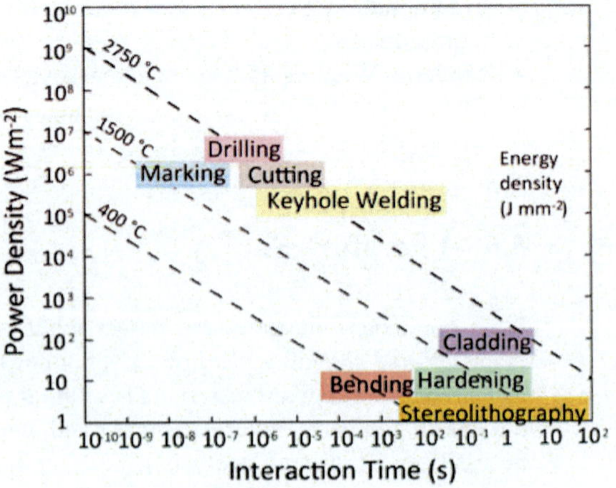

Fig. 5.7 Power density versus interaction time for various laser processes [8]

Liao [9] studied the impact of process parameters in Nd:YAG laser spot welding of SS sheets. Power input was set in the range of 0.6–1.2 J with an incident angle 30–75°. They concluded that the weld spot geometry is controlled by the input power and the lead angle of the laser. The size of the welded spot is characterized by penetration depth (PD), bead length (BL), and bead width (BW) in the curves shown in Figs. 5.8 and 5.9.

The characteristic curves indicate that the penetration depth and weld width increase with the incident angle while the bead length decreases (Fig. 5.8). The aspect ratio was observed to be independent of the laser energy (Fig. 5.9a), and the weld spot volume is not sensitive to incident angle (Fig. 5.9b).

Salminen et al. [10] performed experiments to find the optimum combination of laser welding parameters and also laser setups for getting high-quality full penetration welds of S355 EN 10025 structural steel. Thicknesses up to 25 mm have been shown possible in welding with laser power between 5 and 30 kW. For the purpose of testing the different setups, the comparison was done between fiber laser

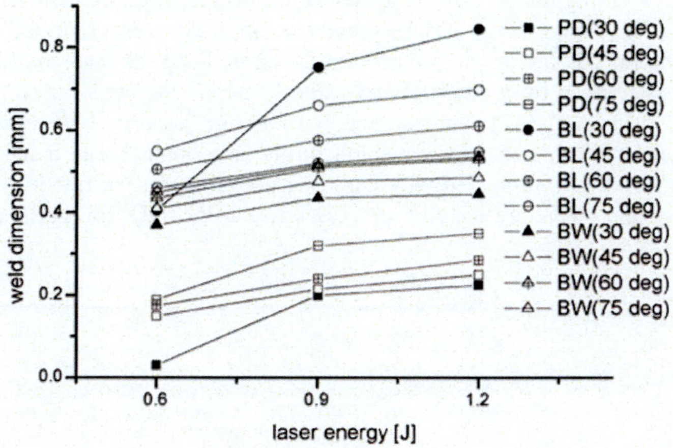

Fig. 5.8 Weld dimensions versus laser energy [9]

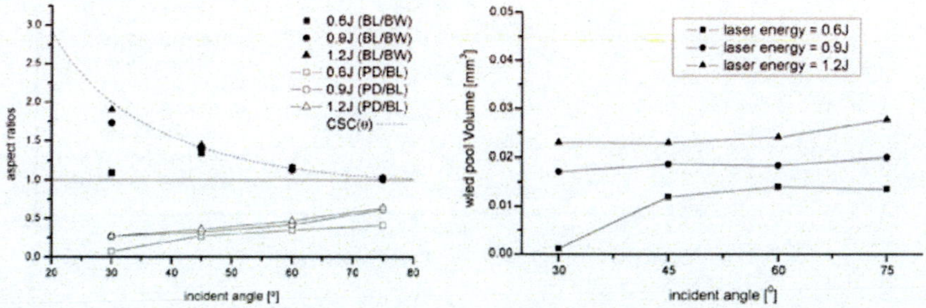

Fig. 5.9 **a** Weld dimensions versus incident angle [9], **b** weld pool volume versus incident angle

and disk laser, different material thicknesses and focal point position. The performance was evaluated in terms of joining speed, weld penetration, and joint edge quality. Results gained with various weld setups show high joining speed with respect to laser power and also show the optimum spot size for achieving the maximum speed (Fig. 5.10).

These curves show significant variation in the values due to different setup arrangements and joint preparations. The laser welding was found to give high productivity with high power fiber.

Caiazzo et al. [11] investigated laser welding of 3 mm Ti-Al-4 V alloy plates for butt configuration weld to get the characterization data of the weld bead geometry. A cognitive methodology was developed based on the employment of ANN for representing nonlinear mappings between a set of variables. This would also enable prediction of optimum process parameters to obtain the required weld geometry. They arrived at a three-dimensional graph shown in Fig. 5.11, enabling visual examination of the pattern recognition performance of ANN.

This three-dimensional graph represents the process parameters predicted by the ANN configuration and the actual experimental process parameters. It is observed that the experimental and the predicted values are closely located in the 3D space. And for this reason, no preferential direction along the three axes is identified, and the deviation from the predicted values is nearly the same for all three parameters.

Kelly et al. [12] demonstrated that a hybrid laser arc butt welding of thin steel panels, suitable for heavy manufacturing environment, can reduce distortion by a factor of 2–4 compared with conventional processes. In hybrid laser arc welding, any one of the arc welding processes like MIG, TIG, or SAW is used with deep

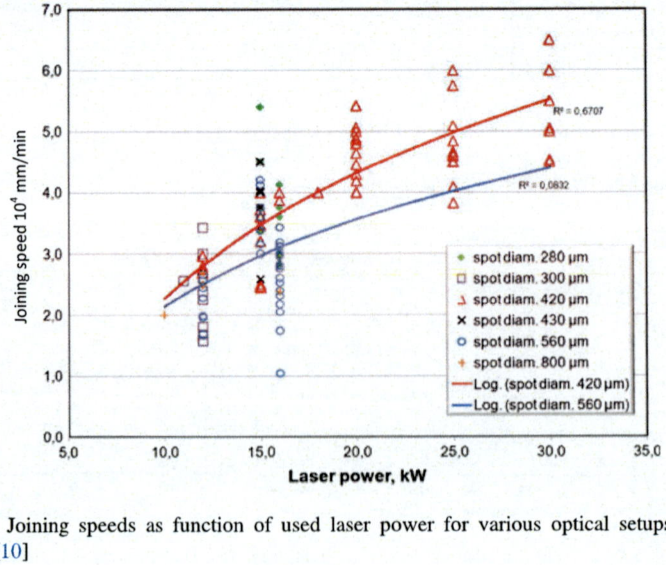

Fig. 5.10 Joining speeds as function of used laser power for various optical setups and spot diameter [10]

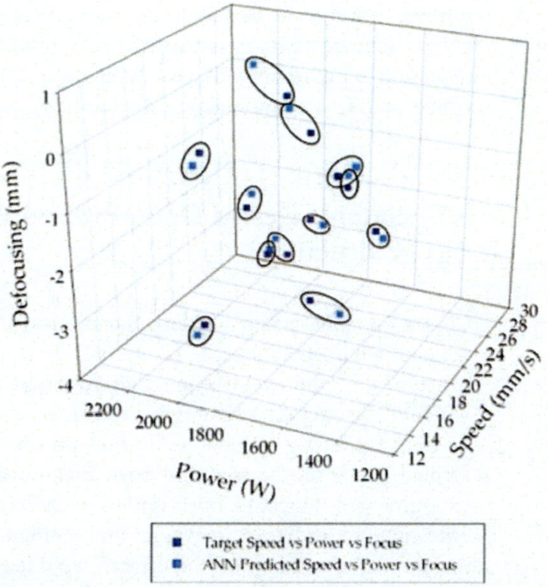

Fig. 5.11 Experimental and ANN estimated values for different experimental conditions [11]

penetration laser welding. The experiments were performed using GMAW with laser welding (Fig. 5.12).

Nayak et al. [13] explored the laser welded blanks (LWB) technique of laser welding for advanced high-strength steels (AHSS) to understand the changes that occur during laser welding. Typical AHSS families used in automotive construction, for example, dual-phase (DP) steel, transformation-induced plasticity (TRIP) steel are characterized by yield strengths and ultimate tensile strengths higher than 300 and 600 MPa. LWBs are composed of two or more sheets of similar or dissimilar materials, thicknesses, and/or coating types welded together, which are

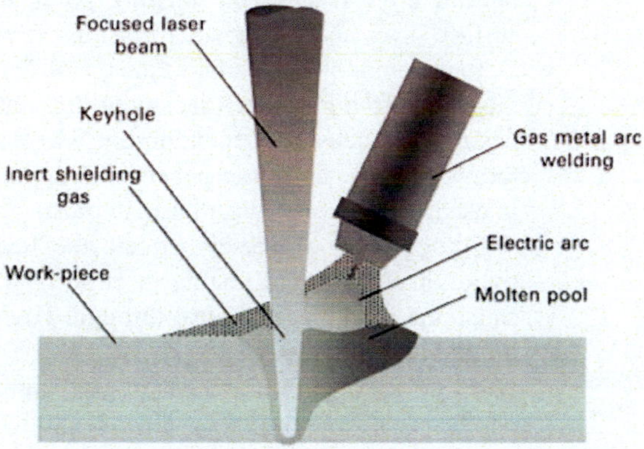

Fig. 5.12 Hybrid laser arc welding [12]

formed to fabricate three-dimensional automotive body parts. The advantages of LWBs include weight reduction, cost minimization, material usage, and scrap reduction with improved part integration. This welding arrangement for AHSS would enable the automakers to effectively design automotive parts.

5.7 Sample Design Data, Process Parameters, and Design Calculations

Choice of appropriate welding technique for a defined welding technique for a defined weld specification is critical as it dictates the quality and mechanical properties of the weldments. Subsequently, the creation of optimal parametric window for operating ranges is crucial to accomplish desired weld efficiency.

Conventionally, selection of weld process operating parametric ranges is performed based on the trial and error method that chiefly rests on the discretion of experimenter, machine operator, or technician. The welds are examined for its efficiency in order to arrive at the optimal parametric ranges that need to be maintained to accomplish the desired weld features. However, this procedure incurs huge cost and is time consuming and exhaustive.

Researchers have reported/proposed a few alternate schemes for various welding processes to minimize this herculean task of parametric range estimations through trial and error method. This chapter documents comprehensively the instruction sets/procedures with empirical equations and formulae to be adopted for forecasting optimal parameter values/ranges for laser welding process.

The critical process parameters to be considered in LBW are [15]

- Power density—Range: 10^4–10^6 W/cm^2 is responsible for surface layer to be heated up to the boiling point or to the point of vaporization. With higher power densities, the time required is in the microsecond range, and the time required is larger for lower power densities. Thus, at higher power densities, large amount of vaporization results in small time duration. With lower power densities, the bottom layer reaches the melting point, even before the vaporization of the surface layer and thus results in a fusion weld.
- Defocus amount—A certain amount of defocusing is required to avoid evaporation due to high power density at the center of the focus. It can be positive or negative defocusing, depending on whether the focal plane is above or below the workpiece. For large penetration depth requirements—negative defocusing is used, and for welding thin materials—positive defocusing is used.
- Welding speed—Inversely affects the heat input per unit time. Lower than optimum weld speed results in large heat input and in its extreme case may cause the workpiece to burn through. Higher weld speed results in lower heat input and inefficient welds.
- Laser pulse waveform and laser pulse width—Impact of the laser beam welding on the weld is dictated by the reflectivity of the metal surface. With a high

intensity laser beam, around 80% of the energy is reflected by the metals. With pulse waveforms, the metal reflectivity changes significantly and the absorption of the laser energy is enhanced.

Laser pulse width is of significance in estimating the cost and volume of processing equipment and also controls the HAZ in the weld process.

- Incident angle—This angle affects the shape and characteristic length of the welded spot. The weld depth to bead length ration increases with the incident angle, whereas the bead length to bead width ration decreases with the increasing incident angle.

5.7.1 Design Sequence in Welding

The design sequence in welding involves the following sections:

- Selection of weld profile for a particular application:
  - Deep, narrow profile: This profile will have lower HAZ and lower tolerance to fit up. The fiber laser diameter should be around its lower limit for deep weld profile.
  - Shallow, wider profile: This profile will have larger HAZ, better tolerance to fit up, and will have greater distortions.

Required weld productivity:

- Longer welds wherein speed is the key factor.
- Short welds, where positioning and clamping of the workpiece is vital.
- Penetration depth decides the weld strength.

Above design aspects are then transformed into the process parameters.

- Long welds will require continuous weld process at high weld speed.
- Short welds may require lesser heat input and pulsed weld process.
- Frequency will depend upon the energy per unit length.
- The laser beam size and the weld tool speed together affect the weld width and are inversely related. For narrow weld, bigger beam size should be used with high weld speeds.
- Thereafter, laser power or current is decided depending upon the required laser control mechanism.

Weld penetration depth is the significant response parameter that further determines the weld resistance length at the interface and thus helps in assessing the shearing strength of the weld joint in overlap configuration.

For accurate prediction of the weld penetration depth directly from the weld input parameters without resorting to the trial mechanism, various analytical and optimization techniques can be implemented that involve defining the response factors by developing mathematical and energy-based models. Use of optimization methods would also have an advantage of giving alternative ideal welding parameter combinations.

Understanding the material properties and behavior is important prior to welding. Estimation of the parametric ranges of the weld process requires meeting the specifications of the weld with consideration of the material capabilities. This section covers various approaches to achieve an efficient weld.

5.7.2 Approach I: Heat Affected Zone Considerations

Estimation of parametric ranges is with respect to the acceptable HAZ for any application [25].

The weld cross section is assumed to be similar to the beam shape for the consideration of Gaussian beam profile while conduction being the mode of heat transfer.

In one general position of the beam, the molten pool is assumed to have semicircular area with its diameter same as the weld width (W), i.e.,

$$\text{Spot Diameter or Molten pool Diameter } \phi = \text{Weld Width } W$$

To simulate the continuous interaction of the laser beam with the weld surface, the irradiated area is considered to be rectangular.

Length L is considered to be half of the weld width, i.e.,

$$\text{Length of Irradiation } L = \frac{W}{2} : W - \text{weld width}$$

For the prediction of the weld penetration depth, the weld cross section is assumed to be a triangle and thus the molten pool volume will be a prism as shown in Fig. 5.13.

Heat treatment can also be used to alter the mechanical and metallurgical properties of the HAZ which depend on the material properties as well [25].

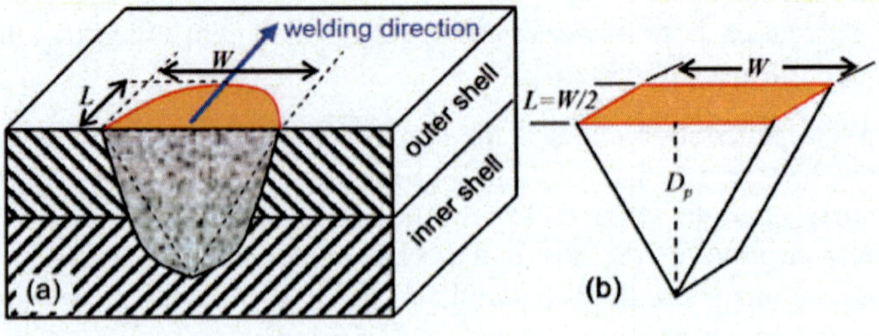

Fig. 5.13 Weld cross section and assumed molten pool volume [28]

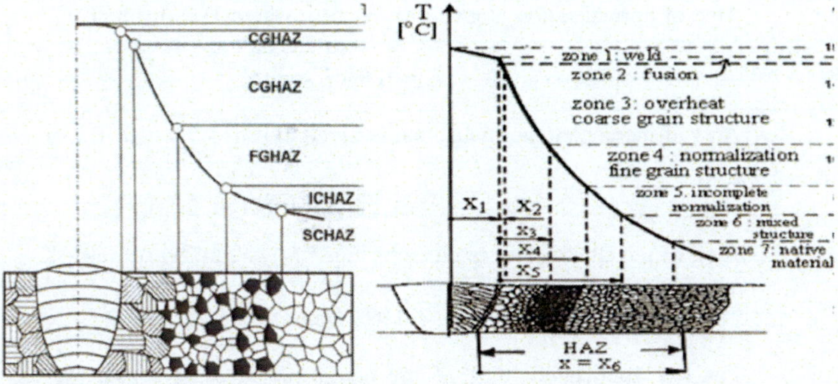

Fig. 5.14 Structural areas of HAZ [28]

The HAZ is a non-homogenous zone and is further characterized into: coarse grain, refined grain, intercritical, subcritical, and partly liquid regions shown in Fig. 5.14, which are formed depending on the cooling rates experienced by the metal [28].

Considering the melting point of the weld metal to be 1,500 °C, the total length of heat affected zone (x) is given by Eq. 5.1

$$x = 1.22\sqrt{kt}; \tag{5.1}$$

This is based upon one-dimensional equation of heat conduction.

k—heat diffusivity coefficient = $0.12 * 10^{-4}\frac{\text{m}^2}{\text{s}}$ for steel, cast steel, and cast iron,
t = time of exposure of weld surface to the temperature = $\frac{L}{v}$
L—extension of irradiated area along weld direction ≈ half of the Weld width half of the laser spot diameter, v—weld speed

For weld velocity of $7.5\frac{\text{m}}{\text{minute}}$ and L = 300 µm: t = 1.2 ms
Thus, length of HAZ x = 0.14 mm
Area of fusion penetration is given by Eq. 5.2, where $T(x, t)$ = 1,480 °C,

$$x_2 = 0.1\sqrt{kt}\,\text{mm} \tag{5.2}$$

Area of overheating is given by Eq. 5.3, where $T(x, t)$ = 1,110 °C,

$$x_3 = 0.51\sqrt{kt}\,\text{mm} \tag{5.3}$$

Area of normalization is given by Eq. 5.4, where $T(x, t)$ = 850 °C,

$$x_4 = 0.834\sqrt{kt}\,\text{mm} \tag{5.4}$$

Area of incomplete normalization is given by Eq. 5.5 where $T(x, t)$ = 750 °C,

$$x_5 = 1.13\sqrt{kt}\,\text{mm} \tag{5.5}$$

Area of mixed structure is given by Eq. 5.6, where $T(x, t)$ = 600 °C,

$$x_6 = x - x_5 = 0.09\sqrt{kt}\,\text{mm} \approx 0.1\sqrt{kt}\,\text{mm} \tag{5.6}$$

For any application, thus, if the acceptable HAZ is considered as the key parameter in the design process, then based on the material properties and the above relations, the weld machine specifications can be determined.

5.7.3 *Approach II: Thermal Gradients with Varying Thickness*

Temperature distribution in the weld material is a function of material thickness.

With an increase in weld heat input or the weld material thickness reduction, the cooling rate of the HAZ area reduces. The cooling rate is proportional to the weld velocity, material thickness, and thermal conductivity [26].

The Gaussian heat source behavior will be distinct for varying thickness of the material [26]. For thick plates, the heat source is considered as a point heat source which involves three-dimensional heat flow as in case of conduction mode welding (Fig. 5.15).

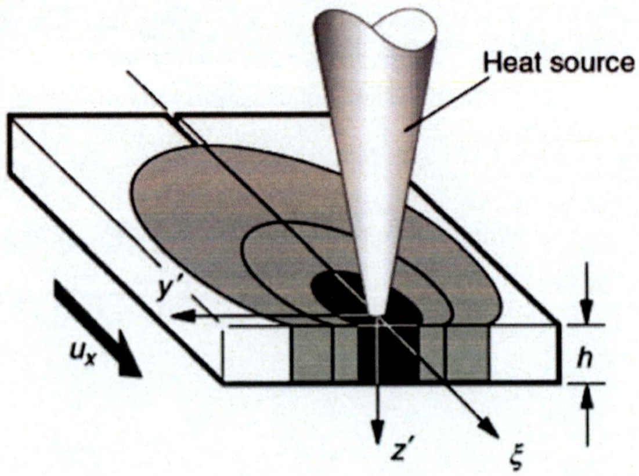

Fig. 5.15 Two-dimensional conduction mode laser beam [26]

In the case of thin plates, the heat source is a line heat source that penetrates through the thickness involving two-dimensional heat flow as in case of keyhole welding. To analyze the heat source as any of the two, the material thickness is to be described. This is specified in terms of factor β_c in Eq. 5.7:

$$\beta_c = h * \frac{\sqrt{[\rho C_p u_x (T - T_0)]}}{\sqrt{Q}} \tag{5.7}$$

Material is considered to be

- Thin when $\beta_c < 0.6$ and
- considered thick when $\beta_c > 0.9$.

The temperature distribution in a plate represented by a family of isotherms is shown for the case of thin and thick plates in Figs. 5.16 and 5.17, respectively [26]. These curves were obtained by solving the Rosenthal relation as mentioned in Eq. 5.8.

$$T - T_0 = \frac{Q}{2\pi k\alpha} \exp\left(-\frac{u_x \zeta}{2k}\right) * K_0 * ((u_x r)/(2k)) \tag{5.8}$$

K_0: Bessel Function of second kind of order zero, $r = \sqrt{\zeta^2 + y^2}$ heat sources radius.

The isotherms give the observation that preheating or increasing the heat input increases the size of the isotherm, thus causes widening of the fusion zone and the HAZ.

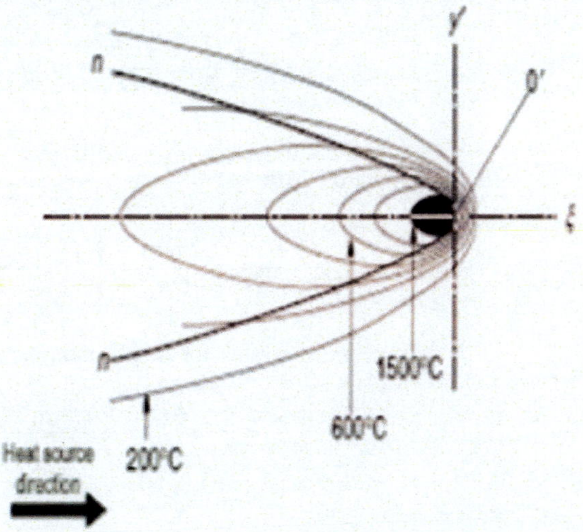

Fig. 5.16 Temperature distribution for thin plate [26]

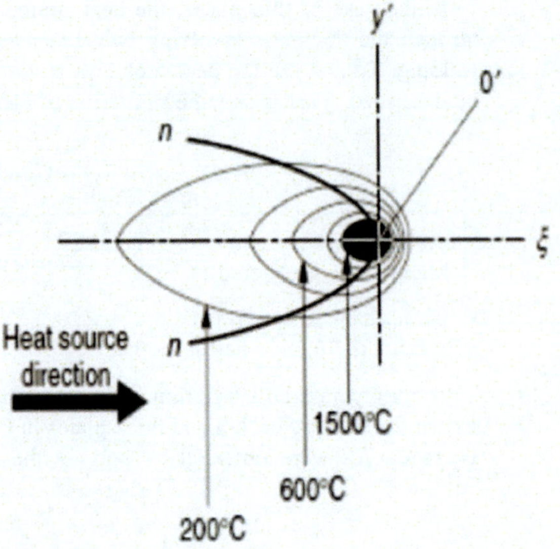

Fig. 5.17 Temperature distribution for thin plate [26]

Also, thinner plate results in a greater HAZ as compared to a thicker plate and high temperature gradient are observed in thicker plate.

The curve n-n is the locus of points that reach their peak temperatures at the same instant of time. Its shape depends upon the transverse speed and thermal diffusivity of the materials.

The peak temperature T_p attained for a thin plate assuming a line source is given by Eq. 5.9:

$$\frac{1}{T_p - T_0} = \frac{\sqrt{2\pi e}}{Q} * \rho C_p \mu_x Y_{HAZ} + \frac{1}{T_m - T_0} \tag{5.9}$$

Whereas, for a thick plate, assuming point source, the peak temperature can be found using Eq. 5.10:

$$\frac{1}{T_p - T_0} = \frac{2\pi k \alpha e}{Q\mu_x} * \left\{ 2 + \left[\frac{\mu_x Y_{HAZ}}{2\alpha} \right]^2 \right\} + \frac{1}{T_m - T_0} \tag{5.10}$$

Also, the cooling rates for thick materials are given by Eq. 5.11:

$$\frac{\delta T}{T} = -\frac{2\pi a \mu_x}{Q} (T - T_0)^2 \tag{5.11}$$

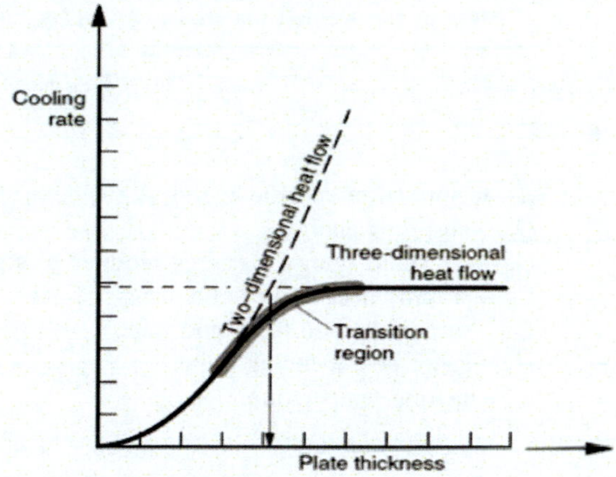

Fig. 5.18 Cooling rates with variation in plate thickness [27]

And for thin plates, the cooling rate is given by Eq. 5.12:

$$\frac{\delta T}{T} = -\frac{2\pi a\rho C_p(\mu_x h)^2}{Q^2}(T - T_0)^3 \quad (5.12)$$

Representative regions of cooling rates are shown Fig. 5.18.

Laser welding process has lower heat input requirement compared to other arc welding processes since the concentrated melting happens over a small region, making it a high intensity process. The low heat input results in a narrow HAZ.

Thus, knowing the material properties, thickness, and the weld parameters of velocity and spot diameter, the geometry of the HAZ can be determined. Based on the permissible HAZ limit for an application, the appropriate laser source and the material can be chosen.

5.7.4 Approach III: Use of Energy-Based Model for Weld Depth Determination

Consider the case of Nd:YAG laser welding in overlap joint configuration on a ferrite stainless steel SS304 [27].

Penetration depth is linearly dependent on the input energy density, and this parameter helps to determine the weld resistance length at the interface. This analysis assumes conduction dominated welding. The weld resistance length further helps in determining the shearing strength in the weld.

Thermophysical properties of the intended material and the process parameters can be used to determine the weld penetration depth. Heat transfer with convection and radiation is neglected in this analysis. Heat source is assumed to be Gaussian rod type.

Mass of the welded material is given by Eq. 5.13 [27].

$$m = K(Q_{ab} - Q_{th}) \tag{5.13}$$

K—reciprocal of specific energy (kg/J) defined in Eq. 5.14,
Q_{ab}—absorbed energy,
Q_{th}—threshold energy = energy required to start the melting process,
C_{ps}—specific heat in the solid state,
C_{pl}—specific heat in the liquid state,
H_m—latent heat mending point,
T_m—melting temperature.

$$K = \frac{1}{\left[C_{ps}(T_m - T_{amb}) + H_m + C_{pl}(T_{max} - T_m)\right]} \frac{\text{Kg}}{\text{J}} \tag{5.14}$$

T_{amb}—ambient temperature, andT_{max}—maximum temperature attained in the weld process (determined as part of experiment).

Energy absorbed by the material is given by Eq. 5.15

$$\text{Absorbed Energy } Q_{ab} = A_c * P_{in} * \Delta t \tag{5.15}$$

where A_c—absorptivity = 0.31 at the wavelength of 1.06 μm, and Δt = irradiation time defined in Eq. 5.16

$$\Delta t = \frac{L}{v} = \frac{\text{extension of irradiated area along weld direction}}{\text{weld speed}} \tag{5.16}$$

Material density is given by Eq. 5.17

$$\rho = \frac{m}{\{\frac{1}{2} W * D_p * L\}}) \tag{5.17}$$

Using Eqs. (5.16) and (5.17) in (5.14) to get the penetration depth in Eq. 5.18 or 5.19:

$$\frac{1}{2} W * L * D_p * \rho = K\left(\frac{A_C P_{in} L}{v} - Q_{th}\right)$$

$$D_p = \frac{2K}{WL\rho}(A_c P_{in} L - Q_{th}) \tag{5.18}$$

$$D_p = \frac{2KA_c}{\rho} * \frac{P_{in}}{vW} - \frac{4KQ_{th}}{\rho W^2} \text{ using } L = \frac{W}{2} \text{ (as considered in Approach I)} \tag{5.19}$$

Q_{th} = minimum heat required to initiate the melting process and can thus be determined by Eq. 5.20 for the situation when penetration depth is zero.

Because of the minimum value of heat Q_{th}, the width of the weld W is taken same as the spot diameter, i.e.,

spot diameter ϕ = weld width W − (as considered in Approach I, refer Fig. 5.13)

$$Q_{th} = \frac{A_c P_{in} W}{2v} = \frac{A_c}{2} * \frac{P_{in}}{v} * \phi_{spot} \tag{5.20}$$

Energy density can be written as Eq. 5.21 and can be modified to get the energy per unit length with Eq. 5.22

$$\mathrm{ED} = \frac{P_{in}}{v * \phi_{spot}} \frac{\mathrm{J}}{\mathrm{mm}^2} \tag{5.21}$$

$$\Rightarrow \frac{P_{in}}{v} = \mathrm{ED} * \phi_{spot} \tag{5.22}$$

Using (5.22) in (5.21),

$$Q_{th} = \frac{A_c}{2} * \mathrm{ED}_{th} * \phi^2_{spot} \Rightarrow \frac{2Q_{th}}{\phi^2_{spot}} = A_c.\mathrm{ED}_{th} \tag{5.23}$$

Equation 5.23 gives the threshold heat in terms of absorptivity and threshold energy density.

ED_{th} is the energy density threshold ratio representing the start of the melt process. This must be determined experimentally and depends on the laser weld setup and the material type.

Using (5.22) and (5.23) in (5.19) to get depth of penetration with Eq. 5.24:

$$\begin{aligned} D_p &= \frac{2K}{\rho} * \{A_c * \mathrm{ED} - A_c * \mathrm{ED}_{th}\} \Rightarrow \\ D_p &= \frac{2KA_c}{\rho} * (\mathrm{ED} - \mathrm{ED}_{th})\mathrm{mm} \end{aligned} \tag{5.24}$$

Thus, penetration depth $D_p = K_m$ (Δ energy density)
where

$$K_m - \text{Material constant} = \frac{2KA_c}{\rho} \tag{5.25}$$

Equation 5.24 confirms the linear dependence of penetration depth on the energy density.

Calculations for SS304 Laser weld with a 1.1 KW CW Nd:YAG laser.

The control parameters for the experimental study are considered in the range:

Laser power Pin = 800–1.1 kW,
Weld speed v = 4.5–7.5 m/min and
Focal spot diameter Φ_{spot} = 300–400 mm.

Material parameters

$C_{ps} = 490\frac{J}{Kg{-}K}; C_{pl} = \frac{800\,J}{kg{-}K}; H_m = \frac{600\,J}{Kg}; \rho = 7.9 * 10^{-6}\,kg/mm^3$

And for the considered conditions:

T_m = 1,400 °C—melting temperature of weld steel,
T_{max} = 2,500 °C; ED_{th} = 2.8 J/mm^2—both parameters derived experimentally [27].

Using these values in (5.14) constant $K = 6.43 * 10^{-7}\frac{kg}{J}$.

Using the values of absorptivity A_c = 0.31, material density = $7.9 * \frac{10^3 kg}{m^3}$ and K in Eq. 5.25, material constant K_m = 0.05.

Using the input power, weld velocity and focal spot diameter in Eq. (5.21):

$$P_{in} = 1100\,W, v = 7.5m/min, \phi_{spot} = 300\mu m \Rightarrow ED = 29.33\,J/m^2$$

Using the obtained values in Eq. 5.25: D_p = 0.05 (ED − ED_{th}) = 1.32 mm.

With the experimental variation in the input power and weld velocity, variation in the ratio of weld resistance length to the weld width (reinforcement form factor) with respect to energy density is shown in Fig. 5.19.

This curve indicates a limiting value of energy density at 32 J/mm^2 at which the weld resistance (S) becomes nearly equal to the weld width (W). Further increase in the energy density results in linear increase in the weld penetration depth—causing change in the weld profile from semicircular to parabolic and finally rectangular as shown in Fig. 5.20, but nearly no change in the weld resistance length.

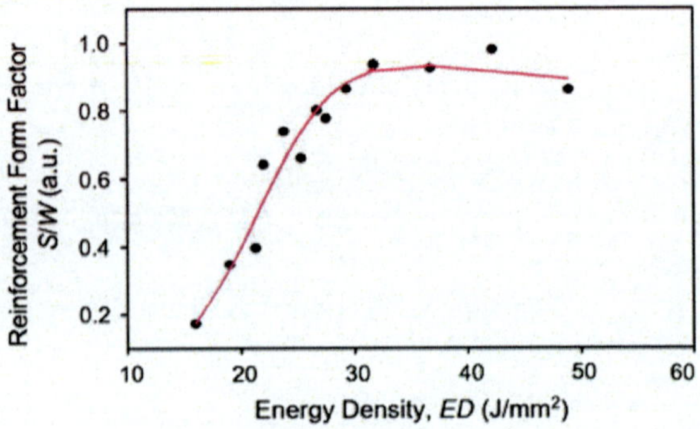

Fig. 5.19 Reinforcement form factor versus energy density [27]

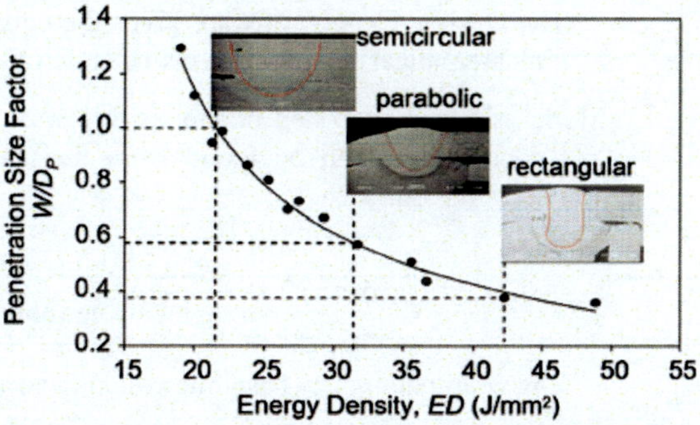

Fig. 5.20 Variation of penetration size factor with energy density [27]

Weld resistance length is the characteristic factor that determines the deformation capacity or shearing strength of the weld. Any change in the energy density after the limiting value will not cause any change in the mechanical strength of the weld, but will only increase the weld penetration depth. Experimental variations also indicate that in the lower range of energy density up to 20 J/mm^2, the penetration depth and the weld resistance length obtained are insufficient for the required weld specifications of any application.

5.7.5 Approach IV: Use of Process Optimization Techniques

Consider the laser weld of AISI416 and AISI440FSE SS in constrained overlap configuration. These materials are used in automotive industry for making the inner and outer shells of a fuel injector [27].

Optimal range of input process parameters: laser power, welding speed and spot diameter is to be found to achieve the required depth-to-width ratio and also the desired resistance length and the shearing force.

Design matrix was created using the Design–Expert software [27].

Input parameters are taken as the laser power (P), welding speed (S), and fiber diameter (D). Industrially recommended values for these parameters as used in automotive industries can be adopted for this study. The ranges of values for parameters are same as in case of the previous case study.

Laser power P = 800–1.1 kW,
weld speed v = 4.5–7.5 m/min and
Focal spot diameter F = 300–400 μm

Mathematical model for this process is developed by implementing the statistical DOE technique of full factorial design. The numerical optimization feature in the design package will find one or more optimum input parameter values to maximize the objective function. The regression models thus developed are tested for significance using ANOVA.

The Design–Expert software gives the equations representing the weld process as a mathematical model in terms of coded factors and actual factors.

Model is based on coded factors as follows:

Coded factors can be found from the actual parameter range and parameter values using Eq. 5.26.

$$\text{Coded Factors} = 2 * \frac{\text{(Actual Value - Average of actual vale)}}{\text{range between low and high actual settings}} \tag{5.26}$$

The weld parameters based on coded factors can be determined using Eqs. 5.27–5.30.

$$\text{Weld Width } W = -221.78917 + 0.26482\,P + 21.185\,v + 1.265\,F \tag{5.27}$$

$$\begin{aligned}\text{Weld Depth } D_{\mathrm{p}} =& 891.94 + 213.06\,P - 250.69\,v - 93.61\,F \\ & - 73.75\,Pv - 26.94\,PF + 30.97\,vF\end{aligned} \tag{5.28}$$

$$\begin{aligned}\text{Resistance Length } S_{\mathrm{L}} =& 423.94 + 104.64\,P - 107.42\,v - 0.5\,F \\ & + 70.5\,Pv + 31.31\,PF - 55.75\,vF\end{aligned} \tag{5.29}$$

$$\begin{aligned}\text{Shear Force } F_{\mathrm{s}} =& 5418.03 + 874.12\,P - 973.25\,v - 489.75\,F \\ & + 642.13\,Pv - 479.75\,vF\end{aligned} \tag{5.30}$$

Model is based on actual factors as follows:

The weld parameters based on actual factors can be determined using Eqs. 5.31–5.33.

$$\text{Weld Width } W = -221.78917 + 0.2648\,P + 21.18\,v + 1.265\,F \tag{5.31}$$

$$\begin{aligned}\text{Weld Depth } D_{\mathrm{p}} =& -995 + 4.64\,P - 0.28\,v - 0.94\,F - 0.33\,Pv - 3.6 \\ & * 10^{-3}\,PF + 0.42\,vF\end{aligned} \tag{5.32}$$

$$\begin{aligned}\text{Resistance Length } S_{\mathrm{L}} =& 1807.2778 - 2.64333\,P - 109.11111\,v + 0.48463\,F \\ & + 0.31333\,Pv + 4.17407 * 10^{-3}\,PF - 0.74333\,vF\end{aligned} \tag{5.33}$$

$$\begin{aligned}\text{Shear Force } F_{\mathrm{s}} =& 10037.31944 - 11.29583\,P - 1121.19444\,v \\ & + 28.515\,F + 2.85389\,Pv - 6.39667\,vF\end{aligned} \tag{5.34}$$

The process parameter optimization can be carried out based on the traditional industrial practices to manufacture low cost and high-quality weld joints by lowering the input power and increasing the weld speed. The criteria for selection of optimum values would be maximum weld speed, weld depth, resistance length and

Table 5.2 Sample calculations for the use of optimization techniques

S. No	P (W)	v (m/min)	F (μm)	W (μm)	Dp (μm)	Sl (μ)	Fs (N)
1	810	4.75	300	472.80	934.095	441.187	6,094.568
2	820	4.9	300	478.62	932.288	443.57	5,899.21
3	840	4.7	300	472.01	1,001.44	460.23	5,794.75
4	800	5.37	300	483.28	828.44	438.87	5,489.61

shear force, and minimum possible input power and weld width. Table 5.2 shows calculations for some random trials.

The actual results are calculated as the average of three measured results for each response. The range of acceptable weld depth being 600–1200 μm. The design software gives the optimum values based on the selected criteria of maximum or minimum value of any of the input parameters or response factors. The optimized data also helps decide the fiber diameter to be around its lower range of 300 μm.

Optimal range of process parameters to obtain acceptable weld depth, resistance, and shearing force is between 800 and 840 W and 4.75–5.37 m/min. For higher values of weld penetration depth, resistance length, and shearing force, respectively, the laser power can be minimized as achieved for the considered criteria. In corollary, higher weld speed can be used to minimize the incurred cost per weld with the required specifications.

This combination of process parameters would cause heat input requirement, reduced energy density input and thus improve upon the weld quality with reduced distortion and defects.

5.7.6 Sample Parameter Values

5.7.6.1 Set I [17]

Experiment conducted for:

- The effects of various laser welding parameters on tensile strength and hardness of the weld
- Stainless Steel 316 having dimensions 200 mm × 125 mm × 2 mm,
  - The welding speed, power, and focal position were chosen as variable parameters based on the previous experimental works and based on accurate responses given by these parameters in response to surface methodology investigation. The level of values chosen for this welding is shown in Table 5.3.

Figure 5.21 shows the effect of process parameters on the ultimate tensile strength (UTS) and hardness.

Table 5.3 Optimum values of the variable parameters

Factor	Level 1	Level 2	Level 3
Power (W)	1,000	1,250	1,500
Focal point (mm)	−3	0	+3
Welding speed (mm/min)	600	700	800

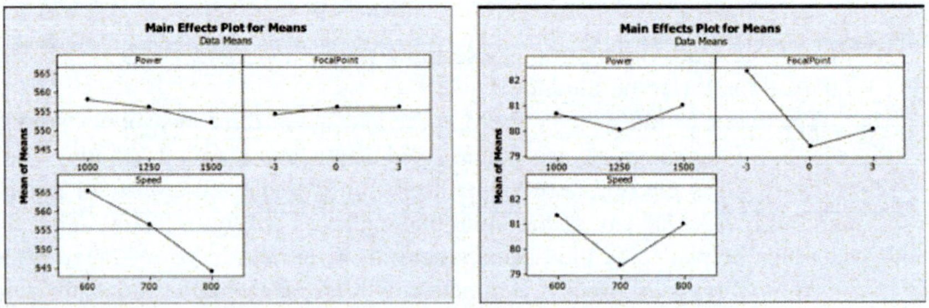

Fig. 5.21 Plots showing effect of process parameters on UTS and hardness [17]

The optimum range of UTS and hardness is obtained with the parameter values given in Table 5.4.

Based on the DOE, the regression equations for hardness and UTS are given in equations

- HW = 212 − 0.0950 P + 0.683 F − 0.197 W − 0.00438 PF − 0.000143 AW + 0.00714 FW
- UTS = 606 + 0.0126 P + 5.50 F − 0.0366 W + 0.00305 PF − 0.000047 PW − 0.0131 FW

Thus, the dependence of UTS and hardness on weld power, focal length, and speed can be summarized as follows:

- The welding speed is observed to affect the tensile strength more than the weld power and the focal point.
- Focal point affects the hardness more than the welding speed and the power.

Table 5.4 UTS and Brinell hardness for given parameter values

Power (W)	Focal point (mm)	Weld speed (mm/min)	Optimum output parameter
1,000	3	600	UTS = 568,22 (MPa)
1,500	−3	600	HRB = 83.55

5.7.6.2 SETII

Laser welding of Austenitic 304 L stainless steel sheet with dimensions: 30 mm * 50 mm * 1.2 mm

304 L [18]

Advantages: low thermal conductivity, high resistance to corrosion, high stability at elevated temperature, and high laser absorptivity.

Experiment 1: Range of welding speed—2 to 10 mm/s
Experiment 2: Range of laser power—300 to 3,500 W

The Nd:YAG laser in multimode condition is used with 4 kW laser beam capacity with process parameters listed in Table 5.5.

Results from the experiments done on the sample show the bead width and penetration depth as functions of weld speed in Fig. 5.22.

The following observations were made:

- The bead width and depth of penetration are inversely proportional to the welding speed
- Depth of penetration is more sensitive to the welding speed than bead width over range of speed selected for the study as the heat input and the interaction time reduce with the change in speed
- In this setup, the weld bead dimensions are more sensitive to the peak power input up to 1,700 W and less sensitive beyond 1,700 W.
- Laser welding machine should not be loaded beyond 98.38% of duty cycle since the depth of penetration reduces drastically beyond this value of duty cycle.

5.7.6.3 SETII

For bead-on welding of AISI 304 stainless steel sheets.

Thickness range—0.1–1.55 mm [19].

Estimation of laser absorptance or coupling efficiency and determination of various process parameters were carried out in both conduction and keyhole welding

Table 5.5 Process parameters chosen for SET II: laser welding of Austenitic 304L stainless steel sheet [18]

Parameters	Values
Gas flow rate	7 l/min
Pulse duration	4 ms
Frequency	25 Hz
Beam angle	90 + −0.5°
Pulse energy	2.76 J
Spot diameter	0.4 mm
Focal distance	150 mm

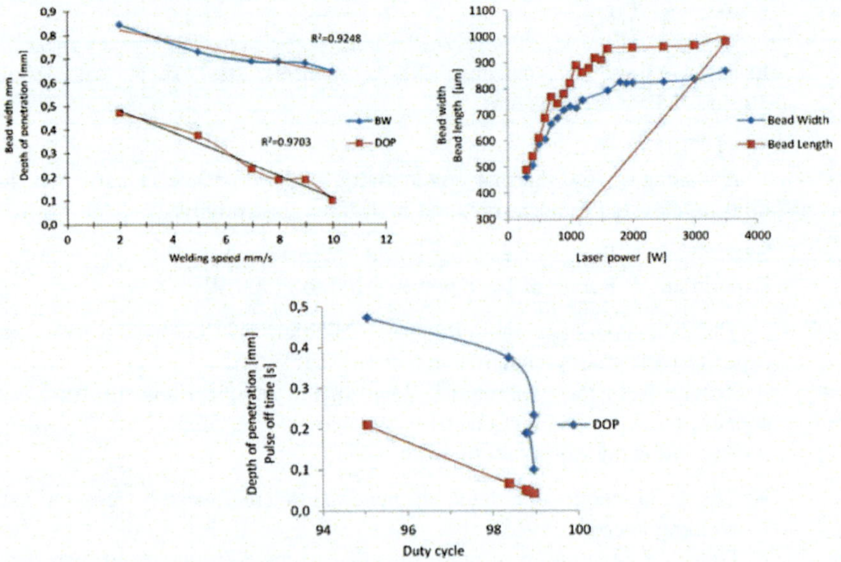

Fig. 5.22 Relation between DOP, pulse off time, and duty cycle [18]

modes. Estimated laser coupling efficiencies in conduction and keyhole welding were about 16% and 65%, respectively. Some of the process parameters were

- Shielding gas—argon flow rate: 20-l/min.
- focal length—100 mm
- spot size—300 micron
- Conduction welding:
  - Sheet thickness—0.1 mm
  - Laser welding power—1,200 W
  - Weld speed—20 m/min
  - Weld width—0.4 mm
- Keyhole welding
  - Sheet thickness—1.5 mm
  - Laser welding power—2250 W
  - Weld speed—3 m/min
  - Weld width—1.3 mm

To improve laser power coupling in conduction welding, different methods such as preoxidization of the surface and use of powder as filler material were implemented. The different ways by preoxidation of metal surface, laser heating in air, chemical passivation, etching of the surface, and with preplaced stainless steel powder as filler material were attempted.

5.7.7 Case Study for Selection of Laser Source

Hermeticity is required for implantable medical devices to protect the microelectronic circuits inside the device from moisture-related failures when the device is implanted in the human body [16].

- Pulsed Nd:YAG laser welding is the most often used hermetic sealing technique in the medical device industry due to the reliability and consistency of seam welds and also to minimize the heat input during laser welding.

Implantable medical devices are designed to be miniature and lightweight so that a thin metallic case is used for hermetic sealing.

- Laser conduction welding mode is more appropriate to achieve the shallow weld of the thin cases.

Fresnel absorption of the laser energy by the target material is not 100% efficient as some energy gets reflected.

An estimate of the absorption capacity, depending on the material properties, should give an idea of the efficiency of the weld.

Absorptivity (%) can be found using Eq. 5.35 arrived as part of this work [16]

$$A = 365.15\sqrt{\frac{\rho_{DC}}{\lambda}}\% \tag{5.35}$$

where ρ_{DC} is the DC resistivity of metals (Ω-m) and λ is the wavelength of laser beams (μm).

For Nd:YAG lasers having a wavelength of 1.06 μm, the absorptivity can be simplified as follows:

$$A_{Nd:YAG} = 354.67\sqrt{\rho_{DC}}\% \tag{5.36}$$

For CO_2 lasers having a wavelength of 10.6 μm, the absorptivity expression simplifies to Eq. 5.37.

$$A_{CO_2} = 112.2\sqrt{\rho_{DC}}\% \tag{5.37}$$

Nd:YAG lasers have a higher absorptivity than CO_2 lasers. Practical absorptivity may be different owing to the presence of vapor and also uneven surface. Oxidation of the metal surfaces helps in improving the absorptivity of the laser beam.

Titanium is an active element and can be oxidized quickly, thus has increased absorptivity and requires reduced laser energy, when exposed to a temperature above 500 ° C in air and the surface turns to be discolored (blue, straw, or purple).

Once the laser beam irradiates on the surface of a substrate and is absorbed by the substrate, the surface temperature increases until melting to form a weld pool. The boundary of liquid and solid phases $X(t)$—melt depth during welding or weld depth after welding—keeps moving down until the laser is turned off.

Melt depth with an assumption of same thermal and physical properties in solid and liquid phases is given by Eq. 5.38:

$$X(t) = \frac{0.16AI(t - t_{\mathrm{m}})}{\rho L} \tag{5.38}$$

A—absorptivity %, I—power density of laser beam, ρ—density, L—latent heat, t—time of start of laser irradiation, and t_{m}—time of melt start at substrate surface

Melting time is given in Eq. 5.39:

$$t_{\mathrm{m}} = \frac{\pi k_{\mathrm{s}}^2 T_{\mathrm{m}}^2}{4\alpha_{\mathrm{s}} A^2 I^2} \tag{5.39}$$

α—thermal diffusivity, k—thermal conductivity for solid phase, and t_{m}—melting temperature.

Thermal diffusivity measures the rate of transfer of heat of a material from hot end to cold end. High value of thermal diffusivity indicates rapid heat conduction.

Melting and vaporizing time for Nd:YAG lasers differ for different power densities.

- Focused laser beam:

Laser beam is focused at a point of interest to achieve high power density. The laser melting process can be simplified as a one-dimensional heat conduction problem with the assumption that the diameter of the laser beam is large enough compared with the regions of interest as shown in Fig. 5.23.

Power density or intensity is given by Eq. 5.40

$$I = \frac{P}{A_1} : P - \text{peak power} \tag{5.40}$$

using area of focused laser beam given by Eq. 5.41

$$A_1 - \text{area of focused laser beam} = \frac{\pi d^2}{4} \tag{5.41}$$

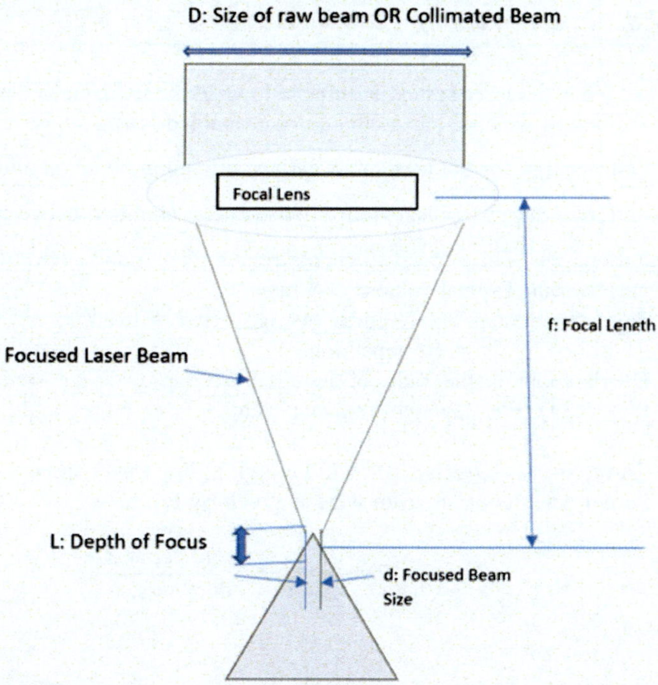

Fig. 5.23 Laser beam focused on the weld metal

Size of focused beam or spot size is given by Eq. 5.42:

$$d = \frac{1.27\lambda f M^2}{D} \tag{5.42}$$

λ—wavelength, f—focal length, D—size of raw beam, and M^2—measure of beam quality indicating power distribution

$$M^2 = 1 \text{ for perfect gaussian distribution}$$

Large value of depth of focus (L) given by Eq. 5.43 is preferred for its impact on process robustness.

$$L = \left(\frac{8\lambda}{\pi}\right)\left(\frac{f}{D}\right)^2 \tag{5.43}$$

5.7.8 Laser Welding Parameters

Pulsed Nd:YAG laser system is suitable to weld the implantable device. Following are the pulsed welding parameters to be examined [16]:

- Laser energy for the hermetic welding of titanium/titanium alloy case: 1–4 J
  - Laser energy—pulse width * pulse energy/duration between two pulses
- Pulsewidth—For a specified energy requirement, power density depends on the pulse width. Typical value = 1–8 ms
- Pulse repetition—The required overlap between two successive spot welds is controlled by the pulse repetition.
- Overlap rate is the ratio of required overlap with the spot weld diameter (Fig. 5.24). For effective hermetic welding, this ratio should be greater than 70%.
- Travel speed—Overlap rate will be high at low travel speeds.
- Partial penetration of seam weld is given by Eq. 5.44

$$\text{Penetration} = \frac{\text{weld depth}, d_1}{\text{thickness}, d_2} \tag{5.44}$$

  - Partial penetration is acceptable for implantable medical devices as they are not subjected to high mechanical stress.

CASE II:

For automotive applications, lightweight metals with better ductility and strength are preferred effectively which reduce the fuel consumption. Advantages of laser welding can be put to use in obtaining the required autogenous components with deep penetration and lesser HAZ. This description gives the optimized design and theory behind the interaction of weld quality and process parameters for DP1000.

Material: DP1000—widely used in automobile and shipping industries for manufacture of frames, wheels having mechanical properties listed in Table 5.6 [21].

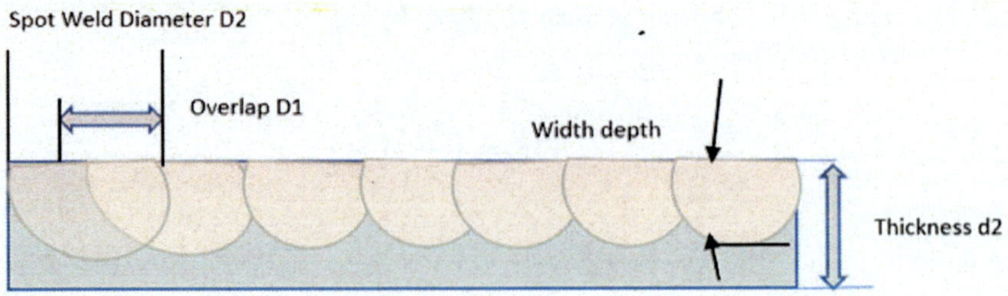

Fig. 5.24 Cross section of pulsed laser seam weld

Table 5.6 Mechanical properties of DP1000 [21]

Elastic modulus (GPa)	Poisson ratio	Yield stress (MPa)	Ultimate stress (MPa)	Elongation (%)	Hardness (HV0.5)
210	0.3	779	1125	9.35	382

Table 5.7 Properties of laser system: SISMA SWA 300

Max. mean power (W)	Wavelength (nm)	Max. peak power (W)	Max. pulse energy (J)	Pulse duration range (ms)	Flash Lamp voltage (V)	Pulse frequency (Hz)
300	1,064	12	100	0.2–25	190–400	0–100

Optimal combination of process parameters is required to achieve the expected penetration depth.

Laser system: SISMA SWA 300 properties of the control system are listed in Table 5.7.

Shielding gas: Argon is supplied to both top and bottom surfaces at 10 l/min.

Spot diameter lies between 0.6 and 2 mm.

Distance between laser exit and sample: 105 mm with optimal resolution.

In the LBW, the weld material gets melted, without it reaching the vaporized state. Thus, the heat energy required for melting will be given by Eq. 5.45:

$$Q = m[c(T_{\mathrm{m}} - T_0) + L_{\mathrm{m}}] \tag{5.45}$$

m—mass, c—specific heat, T_{m}—melting temperature, T_0—initial temperature, and L_{m}—potential heat for melting.

Using the material properties, the value of required heat energy can be found.

Power required for melting is given by energy per unit time (Eq. 5.46)

$$P = \frac{Q}{t} = \rho D h v[c(T_{\mathrm{m}} - T_0) + L_{\mathrm{m}}] \tag{5.46}$$

ρ—density of material, D—laser diameter, h—penetration depth, and v—weld velocity.

Depth of penetration is thus given by Eq. 5.47

$$h = \frac{P}{\{\rho D v[c(T_{\mathrm{m}} - T_0) + L_{\mathrm{m}}]\}} = \frac{KP}{Dv} \tag{5.47}$$

where

$$K = \frac{1}{\{\rho[c(T_m - T_0) + L_m]\}} \tag{5.48}$$

K is the constant associated with surface absorbtion and latent losses given by Eq. 5.48, if H': heat input required for a given length based on material properties = $\frac{P}{v}$.

Then weld depth $h = \frac{KH'}{D}$ mm

Based on the available laser machine specifications of laser power and velocity, heat input per unit length of the weld is given by

$$H = \left(\frac{60W}{v}\right)\frac{\text{kJ}}{\text{mm}} \quad : W - \text{laser power } k \text{ Watts}, v - \text{velocity mm/s}.$$

Comparison of H & H' can help chose the laser machine for the required weld.

Thus, required penetration depth can be found using the power requirement, surface absorption constant, and laser diameter. Penetration depth increases with increase in input power, but is inversely proportional to laser diameter and weld velocity.

The calibration of this particular laser machine is based on the process parameters given below as set of lines in Fig. 5.25 describing the pulse duration and flash lamp voltage [29].

Depending on the pulse duration and the pulse energy, appropriate flashlamp voltage can be estimated or the practical pulse energy can be found using the flash lamp voltage, frequency, and the pulse duration.

The average power density can be calculated with Eq. 5.49:

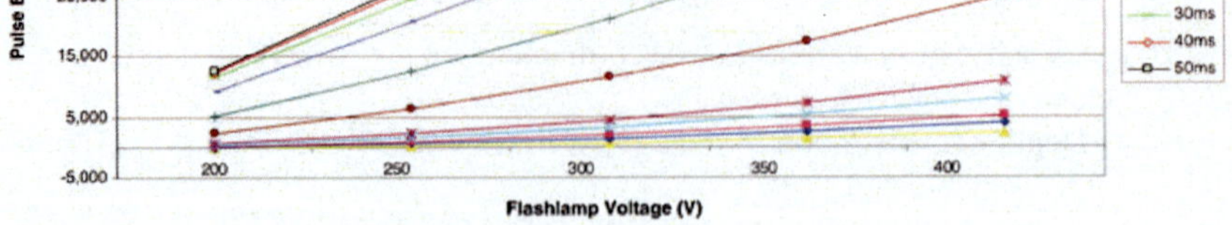

Fig. 5.25 Calibration curve of laser machine [29]

Peak power

$$P_{\text{peak}} = \frac{E}{t}\,\text{kW} \tag{5.49}$$

E—practical pulse energy, *t*—pulse duration.

Pulse energy can be set from laser projecting time and peak power.

Average laser power is defined by Eq. 5.50:

$$P = Ef \tag{5.50}$$

E—practical energy, *f*—repetition rate HZ.

Once the range of peak power, average power, and penetration depth is decided, the weld machine characteristics can be customized and thus selected.

5.7.9 Power Supply for Laser Machine

Operation of flashlamps can be categorized into three steps [20]:

- Triggering and initiation of electrical discharge in the gas: Triggering is the initiation of electrical discharge between the electrodes in the flashlamp, created by the voltage gradient in the path with the externally supplied potential.
  The streamer then fills up the tube as it grows in diameter taking 5.50 μs for its expansion. Amount of charge available from the power supply determined the expansion time. This is the reason for rapid decrease in lamp resistance. This process can be classified as overvoltage, externals, series, and parallel.
- Unconfined discharge—After triggering, current that is flowing in this stage is also minimal because of the high resistance of the gas.
- Wall stabilized plasma stage—For pulsed operations, this is characterized by high value of current flow; for continuous arc lamps, the current value is lower.

Goncz established relation between the voltage and current in flashlamp in Eq. 5.51:

$$V_0 = K_0\sqrt{I} \tag{5.51}$$

K_0—impedance parameter given by Eq. 5.52

$$K_0 = \frac{1.27\left(\frac{p}{450}\right)^{0.2} l}{d_a} \tag{5.52}$$

p—gas fill pressure, *l*—arc length, d_a—arc diameter.

Resistance of the flashlamp during the pulse is given by Eq. 5.53 as a function of time

$$R(t) = 1.28\left(\frac{L}{d}\right)\sqrt{(1/I(t))} \tag{5.53}$$

The current flow varies with time, and the pulse of a typical flash lamp lasts for a few 100 μs (Fig. 5.26). Flashlamp should provide this current pulse. Design of the circuit components is accordingly carried out.

The circuit diagram of the power supply for the flashlamp is given in Fig. 5.27. It provides the following functions.

- Charges a storage capacitor until the flashlamps generate the electrical discharge.
- Provides a high-voltage trigger pulse that initiates the electrical discharge in the gas.
- Controls the flow of current during the discharge.

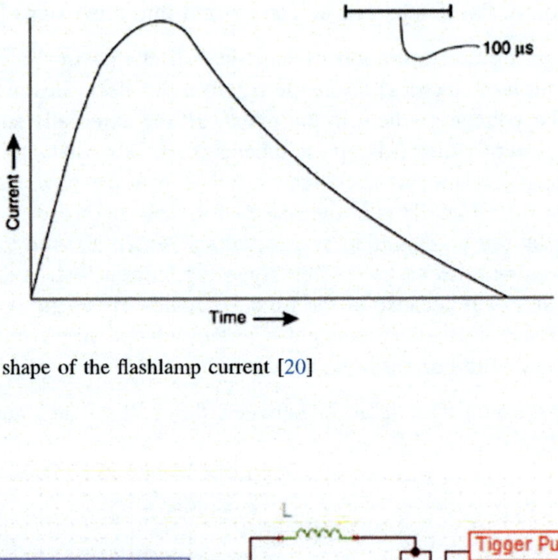

Fig. 5.26 Pulse shape of the flashlamp current [20]

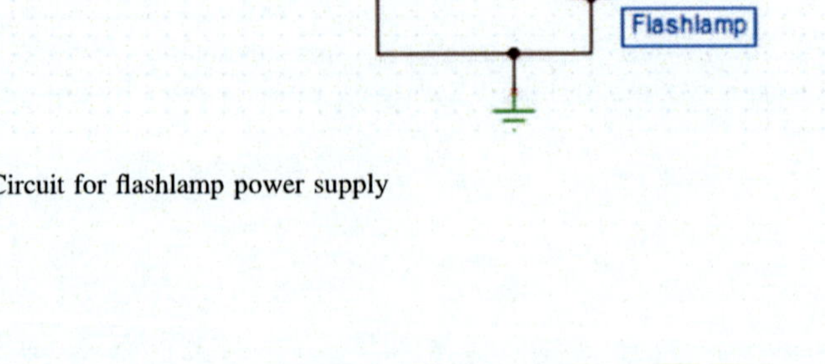

Fig. 5.27 Circuit for flashlamp power supply

5.7.9.1 Charging of Capacitor

The capacitor must get charged up within the pulse duration of the laser. Power supply charging circuit is shown in Fig. 5.28.

Discharge capacitors may be used in the range of 100–400 μF, 500–2000 V.

Energy stored in the capacitors will be given by Eqs. 5.54

$$E = \frac{1}{2} CV^2 \tag{5.54}$$

will be in the range of hundreds of joules, for small lasers.

5.7.9.2 Overvoltage Circuit

In overvoltage triggering circuit (Fig. 5.29), laser discharge is caused due to the breakdown of the gas by the application of initial bias voltage across the gap. The switch used may be a 1,000 volts MOSFET or VFET transistor switch. Specific models of flashlamps have their own trigger voltages. Trigger voltage decreases as the capacitor bias voltage increases. Appropriate polarity of the voltage and the trigger pulse duration should be chosen.

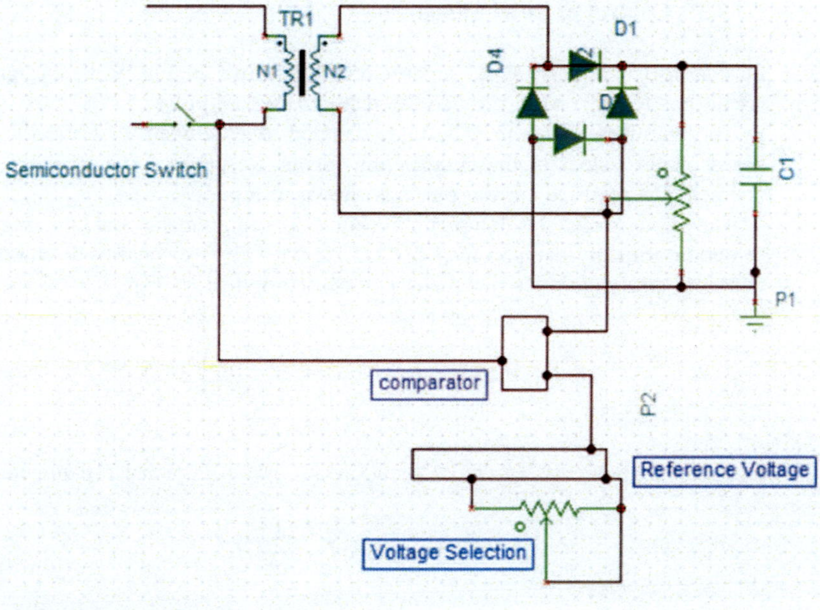

Fig. 5.28 Charging circuit

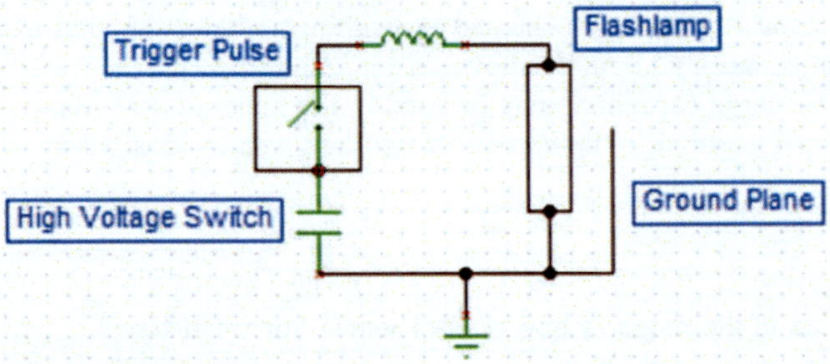

Fig. 5.29 Overvoltage circuit

Trigger pulse must be long enough for the arc to cover up the distance between the electrodes and ensure stable arc length. Recommended value of trigger pulse is 60 nS/cm of arc length.

For an arc length of 2 in (5.08 cm), trigger pulse should be:

$$\text{Trigger pulse} = \frac{60\text{ns}}{\text{cm}} * 5.08\text{cm} = 305\text{ns}$$

5.7.9.3 Control of Pulse Shape

Flashlamp is the basic RLC series circuit as shown in Fig. 5.30. Design of the circuit has to be done to provide the required current pulse.

The circuit behavior may be underdamped, overdamped, or critically damped based on the values of the circuit components. Response for a critically damped circuit is the required current pulse as shown in Fig. 5.31.

The values of the flashlamp circuit (Fig. 5.30) components and charging voltage to obtain critically damped response for known values of energy discharge, pulse duration, and impedance parameter can be determined by Eqs. 5.55–5.57

$$\text{Capacitance } C = \left(0.09\frac{E_0 t_p^2}{K_0^4}\right)^{\frac{1}{3}} F \tag{5.55}$$

E_0—discharge energy, t_p—pulse duration, K_0—lamp impedance parameter

$$\text{Inductance } L = \frac{t_p^2}{9C} H; \tag{5.56}$$

Fig. 5.30 Flashlamp RLC circuit

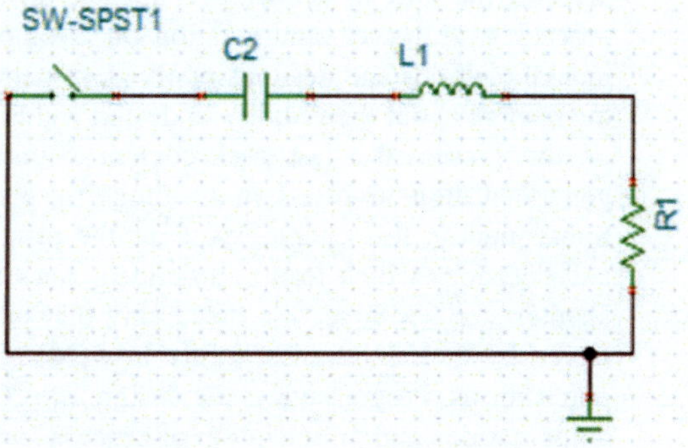

Fig. 5.31 Response of critically damped circuit

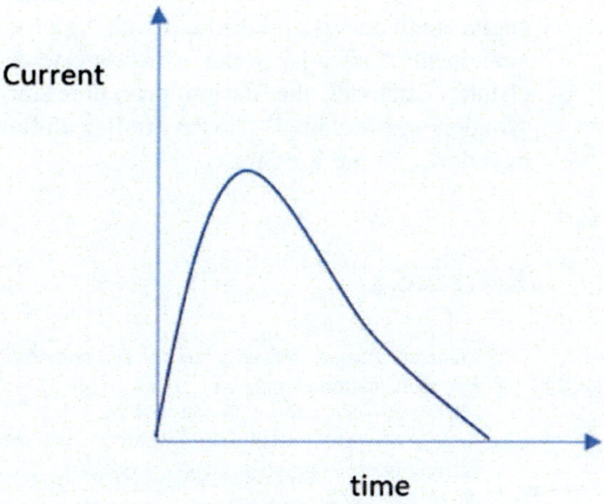

$$\text{Charging Voltage } V = \sqrt{\frac{2E_0}{C}} V \quad (5.57)$$

The above relations describe the laser weld design for a given material and the laser weld machine.

5.8 Conspectus of Design Studies in Laser Beam Welding

The welding engineer's profession is multidisciplinary; weld design engineers are mostly educated and trained in mechanical engineering, electrical engineering, metallurgy, production engineering, or in other disciplines with limited background knowledge on the design. Taking cognizance of this limitation, the book intends to perform the task of knowledge transfer on design aspects for advanced welding processes. However, enumeration on advanced processes by skipping the fundamental methods and designs involved, respectively, will create ambivalence among the readers about their ability to design welds. Hence, Chaps. 3 and 4, taking a note of this, presented a fast track content to understand the foundational substrate of generic design involved in welding irrespective of the joining process chosen. Subsequently, this chapter explains the primary design requirements, procedures, and decisions which lead to a defined characteristic of laser beam weldments. The chapter starts with section discussing the foundational terminologies of lasers followed by their properties, utility in welding, categories, preferable materials, and applications. Progressively, the design aspects are introduced with sample data and computations and optimization of process parameters, respectively, along with case studies for creating authentic learning among readers on laser beam welds. The desired performance of a product relies on materials and methods of fabrication recognized by explicit or implicit design requirements. It is thus imperative for the engineers to understand the laser welding process and design for contributing to the attainment of effective welds which is catered through this chapter. The following chapter discusses the design procedure for another advanced joining process "friction stir welding" chosen from a different weld category to give versatile experience to the learners.

References

1. Branson. Laser welding technical information, Branson. https://www.emerson.com/documents/automation/technical-brief-laser-welding-technical-information-branson-en-us-160146.pdf. Viewed on 13 May 2020
2. Amada Miyachi (2019) Laser welding fundamentals. Amada Weld Tech Inc. https://dev.amadamiyachi.com/wp-content/uploads/2019/12/Laser-Welding-Fundamentals.pdf. Viewed on 13 May 2020
3. Riches, S., 1998. Industrial laser and applications in automotive welding. Lasers in the Automotive Industry, październik
4. Dan Robinson (2009) Laser welding basics, the welder.https://www.thefabricator.com/thewelder/article/laserwelding/laser-welding-basics. Viewed on 14 May 2020
5. Physcis and Radio Electronics. https://www.physics-and-radio-electronics.com/physics/laser/differenttypesoflasers.html. Viewed on 25 May 2020
6. Amey K, Tilekar 1, Nitin K, Kamble, Optimization of laser welding parameters: a review. e-ISSN: 2395-0056
7. Adisa S, Loginova I, Khalil A, Solonin A (2018) Effect of laser welding process parameters and filler metals on the weldability and the mechanical properties of AA7020 aluminium alloy. J Manuf Mater Process 2(2):33
8. Suder WJ, Williams SW (2012) Investigation of the effects of basic laser material interaction parameters in laser welding. J Laser Appl 24(3):032009
9. Liao YC, Yu MH (2007) Effect of laser beam energy and incident angle on the pulse laser welding of stainless-steel thin sheet. J Mater Process Technol 190(1–3):102–108
10. Salminen A, Lappalainen E, Purtonen T (2013) Basic phenomena in high power fiber laser welding of thick section materials. In: Proceedings of the 37th international MATADOR conference. Springer, London, pp 331–336

11. Caiazzo F, Caggiano A (2018) Investigation of laser welding of Ti alloys for cognitive process parameters selection. Materials 11(4):632
12. Kelly SM, Martukanitz RP, Reutzel EW (2011) Minimizing buckling distortion in welding by hybrid laser-arc welding. In: Minimization of welding distortion and buckling. Woodhead Publishing, pp 241–273
13. Nayak SS, Biro E, Zhou Y (2015) Laser welding of advanced high-strength steels (AHSS). In: Welding and joining of advanced high strength steels (AHSS). Woodhead Publishing, pp 71–92
14. Shannon G (2009 Source selection for laser welding. https://www.industrial-lasers.com/welding/article/16484505/source-selection-for-laser-welding. Viewed on 25 May 2020
15. Machine MFG. LASER welding process para meters. https://www.machinemfg.com/laser-welding-process-parameters/. Viewed on 25 May 2020
16. Xie J (2013) Laser hermetic welding of implantable medical devices. In: Joining and assembly of medical materials and devices. Woodhead Publishing, pp 211–235
17. Mazmudar CP, Patel K (2014) Effect of laser welding process parameters on mechanical properties of stainless steel-316. Laser, 1(5)
18. Tadamalle AP, Reddy YP, Ramjee E (2013) Influence of laser welding process parameters on weld pool geometry and duty cycle. Adv Prod Eng Manage 8(1)
19. Nath AK, Sridhar R, Ganesh P, Kaul R (2002) Laser power coupling efficiency in conduction and keyhole welding of austenitic stainless steel. Sadhana 27(3):383–392
20. Zandstra B (2017) PE2BZ datasheet archive. https://pe2bz.philpem.me.uk/Lights/-%20Laser/Info-902-LaserCourse/c04-04/mod04_04.htm. Viewed on 29 Sept 2020
21. Xue X, Pereira AB, Amorim J, Liao J (2017) Effects of pulsed Nd:YAG laser welding parameters on penetration and microstructure characterization of a DP1000 steel butt joint. Metals 7(8):292
22. tvm@2017 (2019) Laser beam welding – equipment, principle, working with advantages and disadvantages. https://www.theweldingmaster.com/laser-beam-welding/. Viewed on 12 Aug 2020
23. Xiao R, Zhang X (2014) Problems and issues in laser beam welding of aluminum–lithium alloys. J Manuf Process 16(2):166–175
24. Amada Weld Tech Inc. (2020) Laser welding fundamentals. https://dev.amadamiyachi.com/wp-content/uploads/2019/12/Laser-Welding-Fundamentals.pdf. Viewed on 04, 2020
25. Satyendra (2016) Heat affected zone and weld metal properties in welding of steels. https://www.ispatguru.com/heat-affected-zone-and-weld-metal-properties-in-welding-of-steels/. Viewed on 30 Sept 2020
26. Kumar KS (2014) Analytical modeling of temperature distribution, peak temperature, cooling rate and thermal cycles in a solid work piece welded by laser welding process. Procedia Mater Sci 6:821–834. lectsuresh25@gmail.com
27. Khan MMA (2012) Laser beam welding of stainless steels
28. Śloderbach Z, Pająk J (2015) Determination of ranges of components of heat affected zone including changes of structure. Arch Metall Mater 60(4):2607–2612
29. Pouquet J, Miranda RM, Quintino L, Williams S (2012) Dissimilar laser welding of NiTi to stainless steel. Int J Adv Manuf Technol 61(1–4):205–212
30. Niggemann D (2018) Mold making technology. https://www.moldmakingtechnology.com/blog/post/choosing-the-right-beam-source-for-your-laser-welding-operation. Viewed on 20 Aug 2020
31. https://en.wikipedia.org/wiki/Laser_beam_welding. Viewed on 12 April 2020

Chapter 6
Friction Stir Welding and Design

Abstract Friction stir welding (FSW) is a solid-state welding process in which the heat generated by friction is utilized to fuse the two weld metals in solidus state. This method uses a non-consumable rotating tool which is moved through the interface of two weld metals causing the materials to soften and thus causing the materials to join. FSW is performed at much lower temperatures than conventional welding techniques. Friction stir welding has proven to be more advantageous as compared to arc welding in producing butt and overlap joints in industrial structures. This chapter on friction stir welding introduces the application of concept of frictional heat used as the heat source in welding process. The subsections cover the requirements of FSW process, working of the FSW machine, compatible materials, applications and significant research advances in this field. The design specifications are the focus point in the last part of this chapter which includes sample design data as a case study for different materials and describes the process parameters and the design sequence with some sample calculations.

6.1 Introduction

Friction stir welding (FSW) has evolved from friction welding which is based on plastic deformation by means of applying pressure and creating a relative movement between two solid bodies. This creates intense local heating which plastically deforms two materials at the point of the interface, bringing them to solidus state. Upon cooling, the weld is formed thus classifying this process as a solid-state welding process.

This is a type of pressure welding that takes advantage of the plastic nature of the materials to be welded. Due to the plastic nature, the deformation caused by certain forces remains even on removal of these forces.

Challenges in fusion welding of aluminum alloys 7075, 7050, 2024 are as follows:

- It reveals cast brittle dendrite structure
- Usually possesses micro porosity

M. Chaturvedi and S. Arungalai Vendan, *Advanced Welding Techniques*,
https://doi.org/10.1007/978-981-33-6621-3_6

- Inferior mechanical and fatigue properties
- Loss of strength in HAZ
- Solidification and liquation cracking
- Loss of alloying elements from the weld pool.

These drawbacks of the fusion processes promoted FSW wherein frictional heat is used for diffusion process.

In FSW, the two workpieces are clamped in a butt joint against a solid support as shown in Fig. 6.1. Welding is carried out with a rotating non-consumable tool similar to a milling cutter. The rotating tool has a pin that penetrates completely through the workpiece. The friction of the rotating tool against the workpiece softens the metal without actually melting it. A collar on the tool prevents the softened metal from being displaced upward that results in smooth underside and the top of the joint. After traversing the workpiece length, the transverse movement is stopped and the tool is retracted from the material leaving behind an exit hole.

Tool Materials—FSW tools of high softening temperature alloys must be capable of retaining their properties at high temperatures. They must be resistant to mechanical wear. The materials which have been observed to perform well as the tool materials are: superabrasive polycrystalline cubic boron nitride (PCBN) and refractory metal tools like tungsten and molybdenum [3] depending on the workpiece material.

Commonly the tool is shaped as shown in Fig. 6.2 with a large diameter shoulder and a small diameter, specially profiled, probe that makes contact first as it is plunged into the joint region. The shape of the tool is designed so that it presses down the weld convexity so that it remains at level with the original surface. Materials having thickness more than 25 mm are most often welded from both the sides. The microstructure of the weld metal is negatively affected by the temperature.

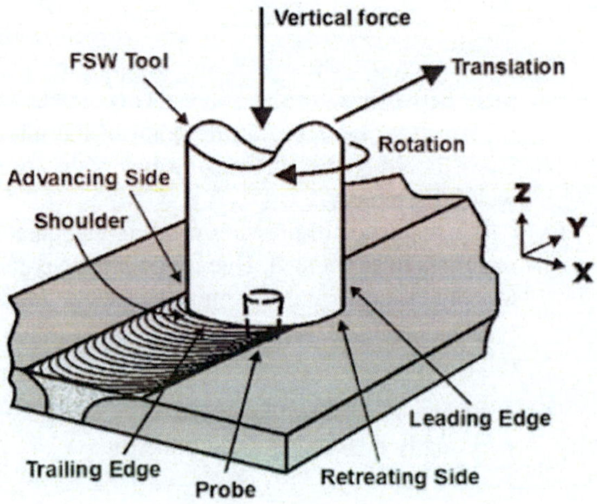

Fig. 6.1 Schematic of friction stir weld process [2]

Fig. 6.2 FSW tool shape [7]

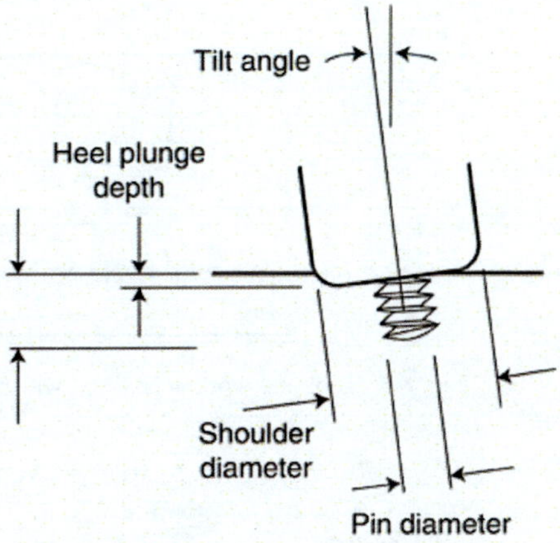

Some forces that act on the tool during welding as shown in Fig. 6.1 are

- Axial or downward force—force required to hold the tool and the weld surface in position.
- Longitudinal or transverse force—force in the direction of the tool motion. It gradually decreases with increase in the temperature around the tool.
- Rotating force whose magnitude will depend upon the downward force, the friction coefficient of the surfaces and the flow strength of the material
- Lateral force- acting normal to the welding direction.

This process is typically divided into four phases:

- Plunge—Rotating tool is forced into the weld line in the workpiece.
- Dwell—When the shoulder of the tool contacts the workpiece surface, tool is kept stationary for a short time. Workpiece gradually heats up, and the surrounding material is softened.
- Weld—Tool is traversed along the weld line to join the two softened parts. Workpiece material gets heated due to the rotation of the forward moving tool. It gets stirred up by the probe such that the material from the two plates merges and creates the weld.
- Retraction—Transverse movement is stopped.

To create more pressure and preventing void formation behind the probe, the tool is often tilted toward the trailing edge of the tool as shown in Fig. 6.3. This "tilt-angle" is typically between 1° and 3°.

The retraction step of FSW poses a disadvantage as when the tool is retracted from the material; it leaves behind an exit hole as shown in Fig. 6.4. The deformed end of the weld due to the exit hole must be removed by sawing or machining as it would be unused.

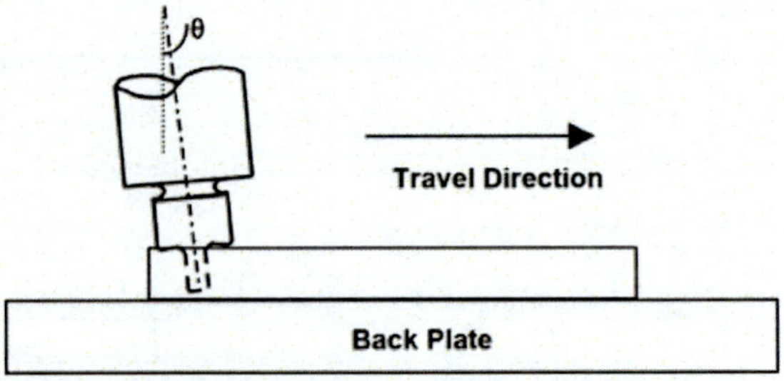

Fig. 6.3 Tool and workpiece tilted with respect to each other [16]

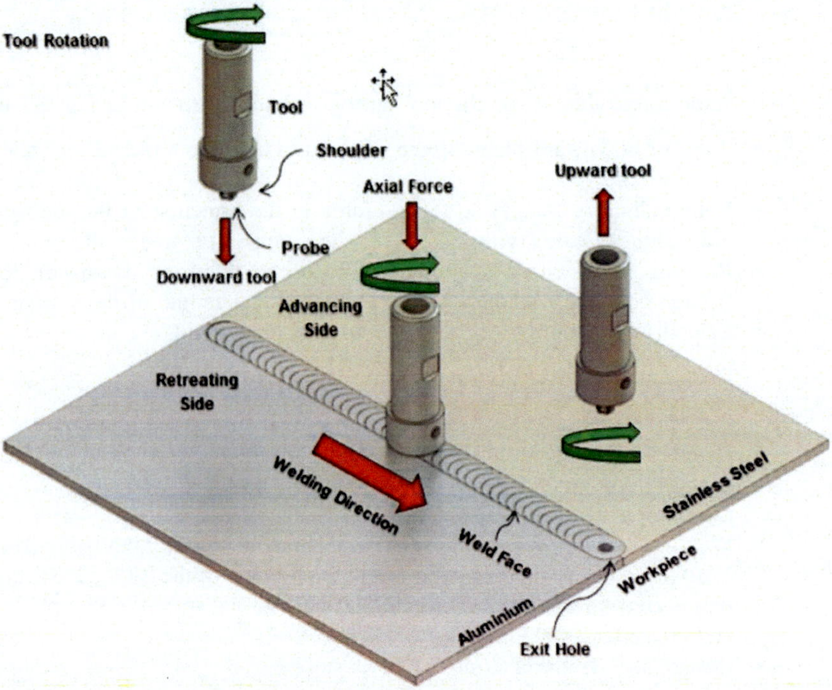

Fig. 6.4 Forces acting in FSW and exit hole [5]

With an increase in the temperature, the stress level needed for the deformation is less, i.e., the yield stress reduces with increase in temperature as depicted in the graph shown in Fig. 6.5.

Advantage in FSW is that preweld cleaning is not as critical as for other welding processes as the rubbing action at the interface breaks up the oxide layer [23].

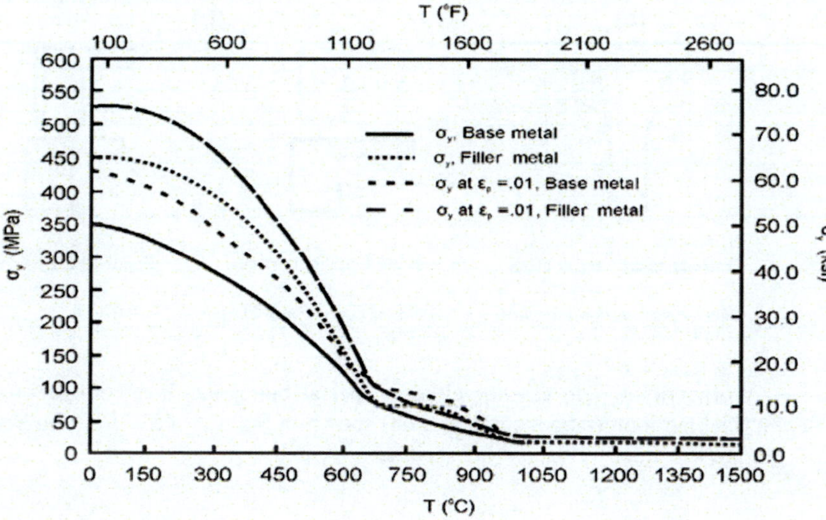

Fig. 6.5 Variation of stress with temperature [24]

FSW is a complicated thermal mechanical coupling process, and the forming quality is mainly affected by the tool shape, rotation speed, traversing speed apart from other process parameters [1].

6.2 Process Applications

FSW process has technical and economic advantage over other processes because of the minimal distortion and high reproducibility.

Some of the examples of automotive applications for FSW which represent the growing use of the technology are: aluminum doors, engine hoods, center tunnel for sports car, suspension links, foldable rear seats, wheel rim from rolled Al 6061-O sheet, and other significant parts.

FSW is also widely used in ship building industry for the manufacture of various products and components. Hollow aluminum panels for the purpose of deep freezing of the fish are one of those components. It is also used for producing prefabricated wide aluminum panels for high-speed ferry boats. The low heat input results in low distortion and reduced thermal stresses.

FSW has been used to prepare spot joints with and without the end keyhole [16]. Spot welds can be either of the butt or lap type. FSW has been also used to prepare T-joints and corner joints. A T-joint could be viewed as a special lap joint and, as such, the notches on either side of the weld are potential crack initiation sites.

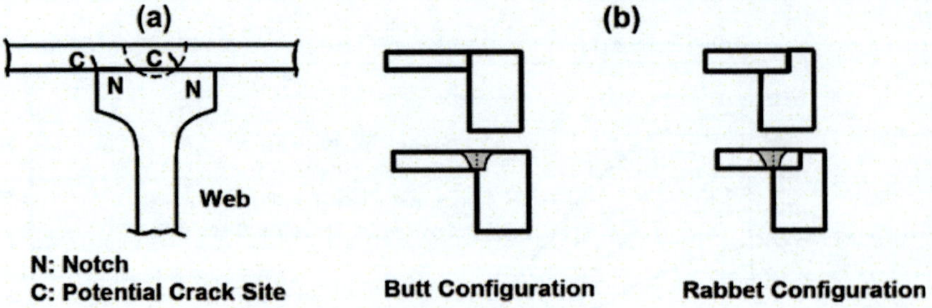

Fig. 6.6 Possible configurations, **a** T-joint, **b** corner joint [16]

A corner joint is in essence either a special butt joint (butt configuration) or a special lap joint (rabbet configuration) shown in Fig. 6.6. FSW has also been used to prepare fillet welds and hem joints.

6.3 Compatible Materials

FSW can weld all aluminum alloys like aluminum–lithium alloys and dissimilar aluminum alloys without the need of a shielding gas. Copper, titanium, magnesium, zinc, and lead can also be welded using this technique. FSW trials have also been done on steel sheets and plates, aluminum-based metal matrix composites, and joining of cast magnesium alloy to extruded aluminum alloy.

6.4 Fundamentals of Friction Stir Weld Process

Main welding parameters which control the FSW process are the rotational speed, traverse speed, axial force on the tool shoulder, the angle of contact between tool and the workpiece [4]. Efficient FSW process depends on the design of the welding tool. Material flow in FSW determines the effectiveness of the joints.

The tool material should be selected such that, depending on the thermal conductivity of material, 95% of the heat gets transferred to the material, only 5% gets transferred to the tool.

Temperature at the weld joint is required to be controlled. Maximum temperature at the weld can be obtained by increasing the angular velocity and radius of the pin within the limits.

FSW occurs at 80–90% of the melting temperature of the weld material.

Total heat generated is a function of the mechanical power delivered to the welding tool. Mechanical power P_m or the total amount of heat generated Q_t depends on angular frequency ω and torque τ given by Eq. 6.1

$$Q_t = P_m = \omega\tau \tag{6.1}$$

The welding tool performs dual movement: Translation (tr) and rotation (rot) and the total amount of generated heat are the sum of translation and rotational-generated heat. The amount of translation heat is negligible and ignored during analysis.

6.4.1 Heat Generation Analysis

The welding contact region (WCR) on the welding tool consists of three areas [17] called the active surfaces of the welding tool (ASWT).

The regions are: probe tip (pt), probe side (ps), and shoulder tip (st) as shown in Fig. 6.7.

Complete welding and all physical processes following it appear on these surfaces or close to them. Total amount of heat generated is the sum of heat generated on every ASWT as shown in Fig. 6.7.

$$Q_{\text{total}} = Q_1 + Q_2 + Q_3$$

$Q_1 = Q_{st}$ = heat generated by the concave shoulder tip
$Q_2 = Q_{ps}$ = heat generated by the pin side
$Q_3 = Q_{pt}$ = heat generated by the pin tip.

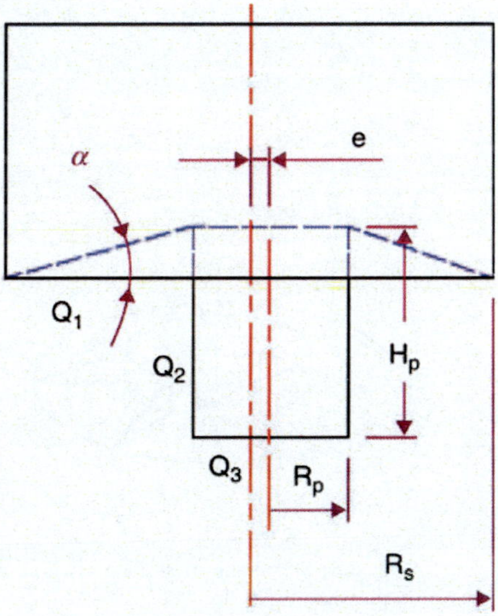

Fig. 6.7 Schematic of FSW tool [17]

The contact surface between eccentric tool and workpiece is given by position and orientation relative to rotation axis as shown in Fig. 6.8. The concave shoulder surface is characterized by the concave angle α, and eccentric cylindrical pin is characterized by eccentricity distance '*e*'. The expressions for each surface area orientation are different, but are based on the general equation for heat generation:

$$dQ = \omega * dM = \omega * r * dF = \omega * r * \tau_{contact} * dA \tag{6.2}$$

where

dQ—heat generated
dM—Torque created
dF—infinitesimal force
dA—infinitesimal surface area
$\tau_{contact}$—contract shear stress.

Heat generation from shoulder surface:
The infinitesimal segment area is given by Eq. 6.3

$$dA_1 = r * d\theta * ds \tag{6.3}$$

is exposed to a uniform contact shear stress $\tau_{contact}$ where r is the distance from the considered area to the center of rotation, ω is the angular velocity, and $r.d\theta$ and ds are the segment dimensions, $ds = dr/\cos\alpha$.

Shoulder heat generation Q_1 is given by Eqs. 6.4 and 6.5:

$$Q_1 = \int_0^{2\pi} \int_{R_p} \omega * r^2 * r_{contact} * d\theta * \frac{dr}{\cos\alpha} \tag{6.4}$$

$$Q_1 = 2\pi * \omega * \tau_{contact} * \frac{R_s^3 - R_p^3}{3 * \cos\alpha} \tag{6.5}$$

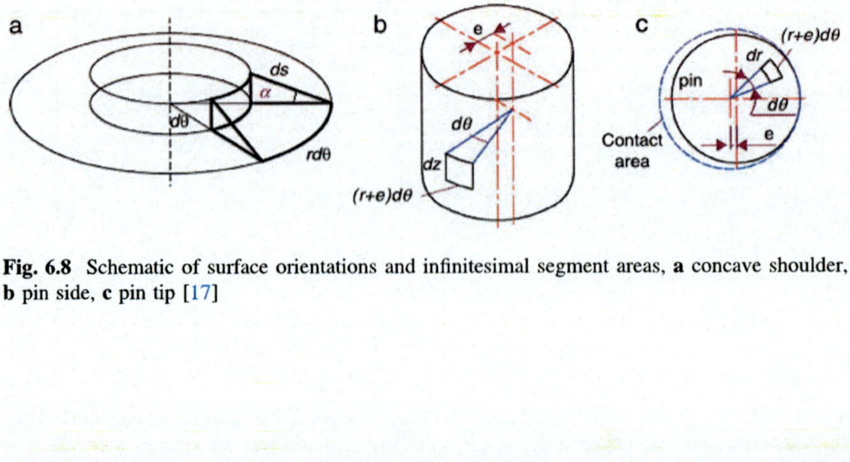

Fig. 6.8 Schematic of surface orientations and infinitesimal segment areas, **a** concave shoulder, **b** pin side, **c** pin tip [17]

Heat generation from pin side surface: The pin consists of eccentric cylindrical surface with a radius of R_p, eccentricity distance e, and pin height H_p.

Heat generated from the pin side Q_2 is given by Eq. 6.6:

$$Q_2 = \int_0^{2\pi}\int_0^{H_p} \omega * (r+e)^2 * \tau_{\text{contact}} * d\theta * dz = 2\pi.\omega * \tau_{\text{contact}} * \left(R_p+e\right)^2 * H_p \quad (6.6)$$

Heat generation from tip surface is given by Eq. 6.7:

$$\begin{aligned} Q_3 &= \int_0^{2\pi}\int_0^{R_p} \omega * (r+e)^2 * \tau_{\text{contact}} * d\theta * dr \\ Q_3 &= \frac{2}{3} * \pi * \omega * \tau_{\text{contact}}\left(\left(R_p+e\right)^3 - e^3\right) \end{aligned} \quad (6.7)$$

Total heat generated is given by:

$$Q_{\text{total}} = \frac{2}{3} * \pi * \omega * \tau_{\text{contact}} * \left[\frac{R_s^3 - R_p^3}{\cos\alpha} + 3H_p * \left(R_p+e\right)^2 + \left(\left(R_p+e\right)^3 - e^3\right)\right] \quad (6.8)$$

In the case of a flat shoulder ($\alpha = 0$) and pin without eccentricity ($e = 0$), the heat generation expression simplifies to Eq. 6.9:

$$Q_{\text{total}} = \frac{2}{3} * \pi * \omega * \tau_{\text{contact}} * \left[R_s^3 + 3H_p * R_P^2\right] \quad (6.9)$$

Energy per unit length of the weld can be found by dividing the total heat generated by the weld speed as in Eq. 6.10.

$$Q_{\text{Energy/Length}} = \frac{2\omega \cdot F \cdot \mu}{3v \cdot R_s^2}\left[\frac{\left(R_s^3 - R_p^3\right)}{\cos\ \alpha} + 3H_p(R_p+e)^2 + \left((R_p+e)^3 - e^3\right)\right] \quad (6.10)$$

The effective energy per unit weld length depends on the transfer efficiency which is defined as the ratio of pin length H_p to the workpiece thickness 't' in Eq. 6.11.

$$Q_{\text{Eff}} = \beta \cdot Q_{\text{Energy/Length}} = (H_p/t) \times Q_{\text{Energy/Length}} \quad (6.11)$$

Direct or indirect heat sources provide the total energy required to weld the workpieces. Process parameters like the tool and weld geometry, weld speed, and material properties control the weld energy.

The empirical relationship between the temperature ratio and the effective energy level that can be applied in case of FSW of aluminum alloys using tool pin with eccentricity is found to be given by Eq. 6.12 [16]:

$$\frac{T_{\max}}{T_s} = 2 * 10^{-4} * Q_{\text{eff}} + \frac{0.5R_p^2}{(R_p + e)^2} \tag{6.12}$$

where

T_max = maximum welding temperature
T_s—solidus temperature

This is characteristic of different aluminum alloys that have approximately the same thermal diffusivity. This model of heat generation can be used to calculate the energy per unit length and the peak temperature at the weld. Other important features of FSW are torque, power needed for welding, and geometry of the stir zone. These parameters should also be predictable with the numerical models of FSW.

6.5 Friction Stir Welding Machine Details

Developments in FSW machinery have been basically concerned with the advancements of welding tools and equipment. Development of new welding tools can be attributed to welding tool design and welding tool material [7].

In the conventional FSW as shown in Fig. 6.9, the workpiece is fixed on to an anvil and the tools move in only one direction to approach the workpiece. This type of tool partially penetrates the workpiece.

Another type of FSW tool known as Bobbin FSW tool is in two parts—upper shoulder and lower shoulder. They are connected by a pin, and the parts rotate together as shown in Fig. 6.10. The two shoulders contain the softened metal from both the sides while generating heat from friction and plastic deformation.

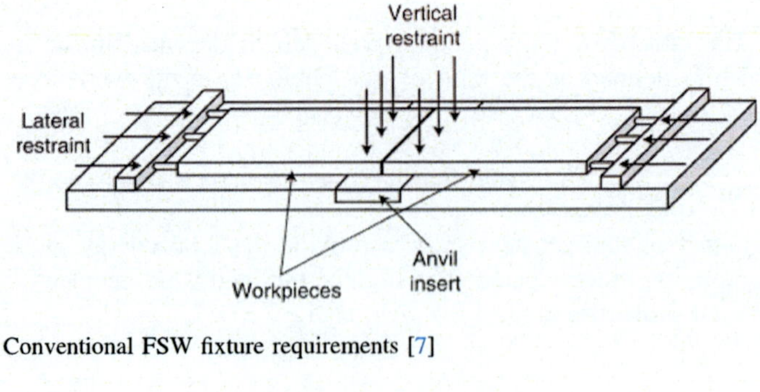

Fig. 6.9 Conventional FSW fixture requirements [7]

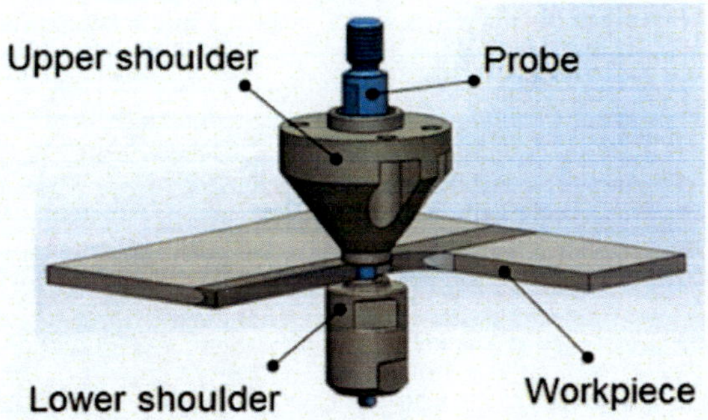

Fig. 6.10 Bobbin FSW tool [5]

The pin serves the purpose of distorting the faying surfaces and generates additional heat to sustain the process. This type of tool completely penetrates the workpiece.

Self-reacting Bobbin tool-spiral grooves/scrolls, as shown in Fig. 6.11, are added to the flat shoulders which act to pull the workpiece material toward the pin. Then, the shoulders are forcibly actuated thus providing for a variable gap. This is allowed for a control over the compressive forces which would be required in case of workpieces having varying thickness along the weld length.

Tapered shoulder tool—This is a fixed gap Bobbin tool as shown in Fig. 6.12 in which the tapered shoulder tool works as a protruding shoulder along with spiral shoulder scrolls. This tool enables to have variable shoulder penetration and also effective shoulder width.

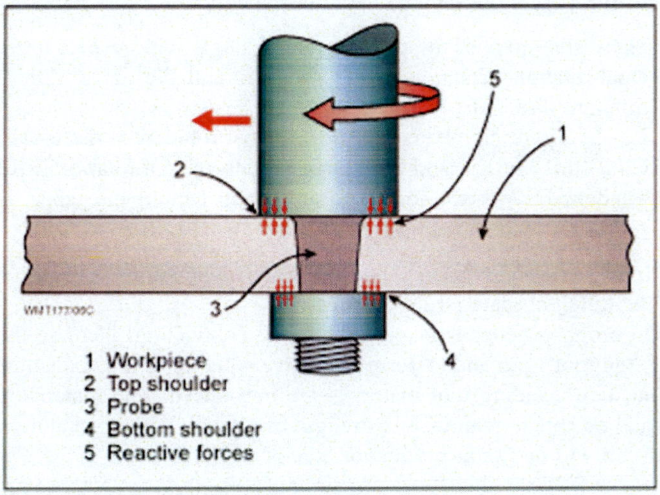

Fig. 6.11 Self-reacting Bobbin tool [22]

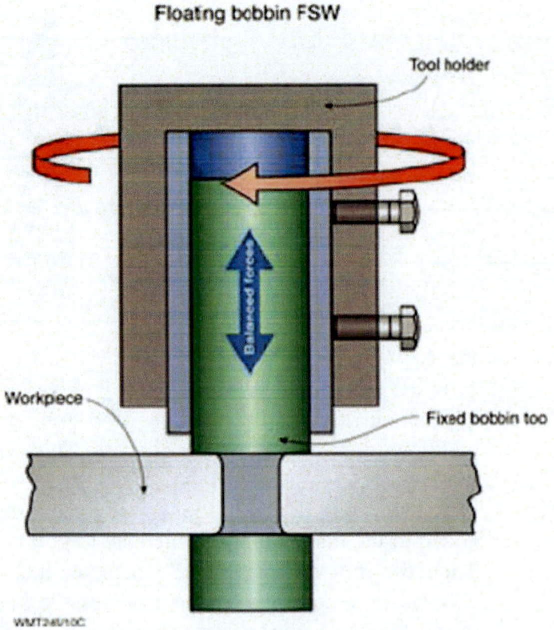

Fig. 6.12 Fixed gap Bobbin tool [22]

Various welding tool features have been introduced over time, and a combination of these features may be used to get a wide variety of conventional and Bobbin FSW tools. Some of the design features are listed in Table 6.1 [7]:

The motion of the welding tool can be varied to achieve desirable effects in weld formation [7]. These include variations such as

- Skew stir—axis of the tool is at some angle with respect to the rotation axis.
- Dual-rotation—rotation of the shoulder and the pin at differing speeds or in different direction
- Re-stir—periodic reversal of direction of rotation of the tool
- Com-stir—vertical and horizontal motion of tool rotation and tool axis orbit
- Tandem FSW—two tools operating one behind the other, like bicycle with pedals for two riders.

Improved weld is formed due to the above variations in the tool motions because of the different material flow patterns.

In order to better manage the thermal conduction, increase the strength of the welding tool's pin and to permit effective welding of materials having high melting point, new welding tool materials are introduced. The material used for welding should be able to maintain its strength at the welding temperature so that maximum tool life and optimum production rate is ensured.

Table 6.1 Design features and effects of the FSW tool

Feature	Intended effect	Examples
Threads on pin	Compression of weld zone against anvil	
Flats	New mode of plastic work, thicker section welding	
Flat pin tip	Improved TMAZ penetration, higher penetration ligaments, better robustness	
Frustum pin profile	Reduced lateral forces, thicker section welding	
Flare pin profile	Wider root profile	
Shoulder scrolls	Elimination of tool tilt requirement, containment of softened workpiece material	
Tapered shoulder	Variable shoulder contact width, variable shoulder penetration	

- Tool Steel such as H13 is used to weld Aluminum alloys
- Nimonic 105 for tool pin and Densimet for the shoulder is used for welding copper.
- Refractory metals such as lanthanated tungsten or tungsten rhenium tools are used for welding Titanium
- Tungsten-based or polycrystalline cubic boron nitride (PCBN) material tools are used for welding steel.

Also, to reduce the wear and to improve the chemical resistance, tool coatings are occasionally used. The microstructural classification employed for FSW, as recommended by The Welding Institute, is shown in Fig. 6.13. The microstructure depends on alloy composition, initial material temperature, welding parameters, tool geometry, and cooling rate.

The weld region is surrounded on each farther side by the parent metal (A) or the base metal which remains unaltered. During welding, the parent metal experiences elevated temperature, but it still exhibits the same properties as the workpiece as in the original condition.

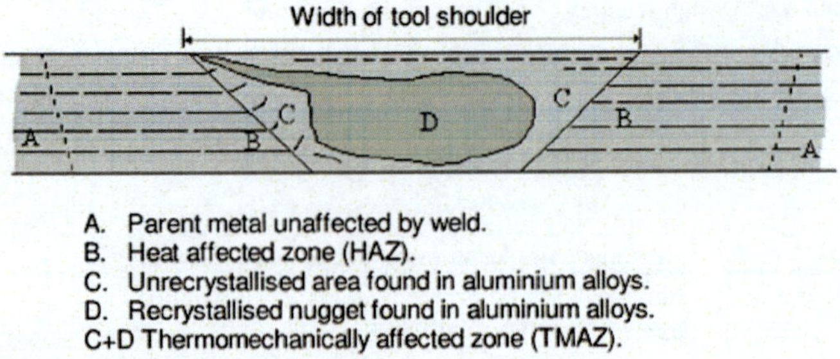

Fig. 6.13 Microstructural classification for FSW product [21]

The next closer region is called the heat affected zone (HAZ)—The HAZ (B) is characterized by more equiaxed grains when compared with the parent plate grains, but no plastic deformation is observed. The properties of this region like strength, ductility, and toughness get altered due to the heating effect, but the original grain structure of the base metal remains unaffected. Corrosion susceptibility also gets affected due to the heating effect.

The thermomechanically affected zone (TMAZ) refers to the plastically deformed material within the joint region. In this region, plastic deformation is observed due to significant heating and by the process forces.

The TMAZ can be further divided into:

- Uncrystallized TMAZ—In this zone, upward pattern of grain deformation is seen around the actual weld nugget and the deformation strain is not sufficient to cause full recrystallization. High density of grains is found in the sub-boundaries.
- Recrystallized TMAZ (Stir Zone or nugget)—This zone experiences the maximum deformation as a consequence of the rotating tool. It exhibits a fine-grained non-uniform microstructure. Due to variation in temperature, the grain size is observed to be different at the top and the bottom surfaces.

The weld zone in FSW being asymmetric in nature results in the zones that have differences in the material flow and temperature ranges, as shown in Table 6.2.

In aluminum alloys, the uncrystallized TMAZ represents a region of low microhardness and increased corrosion susceptibility. It can be of significant size, thus affecting the overall properties of the weld. The high temperature in the HAZ

Table 6.2 Weld zones and their characteristics

Zone	Material flow	Temperature
Nugget	High	High
TMAZ	Low	Medium
HAZ	None	Medium (lower than solidus temperature)

causes change in the material properties due to the recovery of cold work and coarsening of precipitates.

6.6 Literature Survey

FSW is essentially a temperature-dependent process where the microstructure gets negatively altered by the temperature. Even when the weld is run at constant input parameters, the temperature may change over time due to transients and disturbances which can result in unexpected variations in material properties.

Some of the influencing welding parameters are:

- TOOL ROTATION AND TRAVERSE SPEEDS: For best quality weld, it is required to maintain optimum heat for proper diffusion. The heat generated increases with increase in rotation speed and decrease in traverse speed. Extreme high temperature may cause deterioration of weld properties, and the tool may break in case of exposure to cooler environment.
- PLUNGE DEPTH: It is the lowest depth the tool is permitted to penetrate. This depth needs to be adjusted according to the maximum pressure that can be applied. For adequate forging inside the material, the tool should be plunged below the surface. Unrecommended pressure value due to uncontrolled plunge depth may create weld deflections.
- TOOL TILT: For a good weld having elliptical torsion, the rear part of the tool should be lower than the front part. This requires a tool tilt of 2°–4°.
- TOOL DESIGN AND PROPERTIES: Care should be taken in the design and selection of the tool such that it has sufficient strength, toughness, good corrosion resistance, and low thermal conductivity.

Various researchers have investigated various control schemes of FSW by developing different models like first-order plus dead time and heat flow model for the process. Movement of the pin along the three axes and the spindle rotation were considered as the control parameters in those studies. The model predictions of the stir zone geometry, torque, and energy agree well with the corresponding measured values when appropriate values of the heat transfer coefficient and the friction coefficient values are used.

Ross [9] identified the system as FOPDT system and used a PID controller to control the temperature by manipulating the input power. They concluded that power control is best achieved by controlling torque rather than spindle speed. Experimental and numerical results indicate that the temperature of the tool can be approximated within 1 °C by a first-order transfer function with time delay and that the temperature response is dominated by thermocouple location. Effect of changing PID gain for Al is shown in Fig. 6.14. Torque, spindle speed, power and temperature response to step increases in desired temperature are shown in Fig. 6.15.

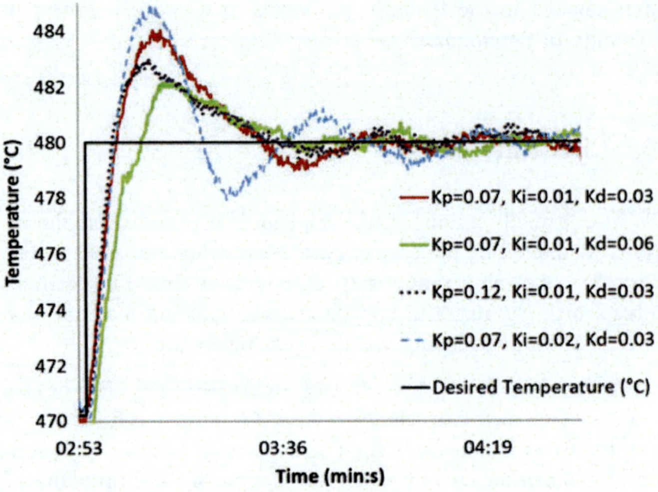

Fig. 6.14 Effect of PID gains on temperature response [9]

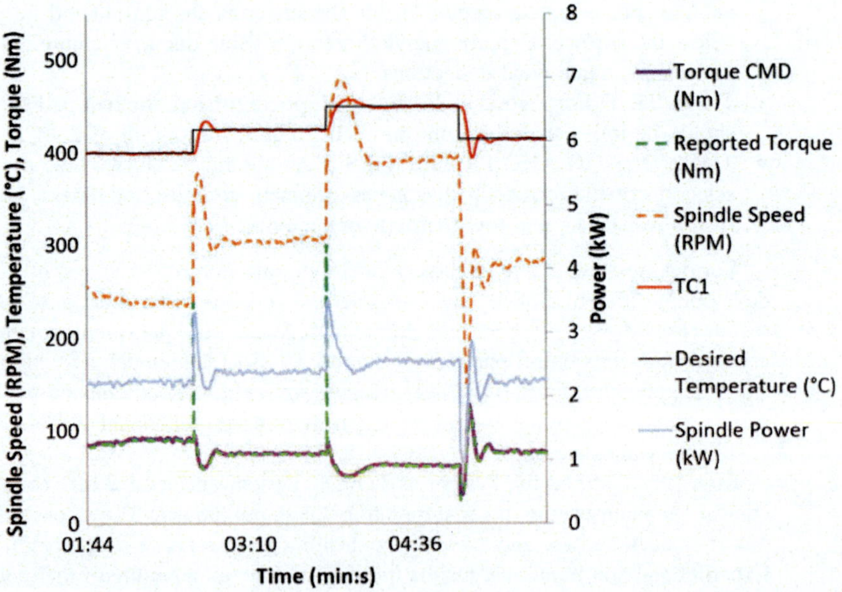

Fig. 6.15 Torque, spindle speed power, and temperature response to step changes in desired temperature [9]

Schmidt et al. [10] investigated the effect of including the tool probe and the material flow in the numerical modeling of the heat flow in FSW assuming that the contact condition at the interface between tool and the workpiece controls the heat transfer mechanisms. Six cases of contact conditions have been introduced based on fully sticking, fully slipping and partial sticking/slipping conditions which were carried out in a model based on thermo-pseudo-mechanical simulation. Three heat sources included—(a) shoulder only heat source, (b) shoulder/probe heat source in which the tool probe is represented as a volume flux, and (c) shoulder probe heat source with no volume flux in the tool/workpiece interface. They found that the convective heat transfer due to material flow greatly affects the temperature fields. The results revealed that the temperature field not only depends on the total heat generation but also on the contact conditions, tool rotational speed, and shear layer thickness.

Arora et al. [8] modeled the FSW process on the basis of the physical laws and parameters of interest like mass, momentum, and energy, and made observations for variations in torque, energy, and size of TMAZ with respect to the weld speed and the rotational speed. They concluded that the energy per unit length is inversely proportional to the welding speed, while torque requirement decreases with increase in rotational speed. This was attributed to the fact that material flows easily at higher temperatures and high strain rates. The size of TMAZ increases slightly with the increase in tool rotational speed. Some of the results are shown in Fig. 6.16.

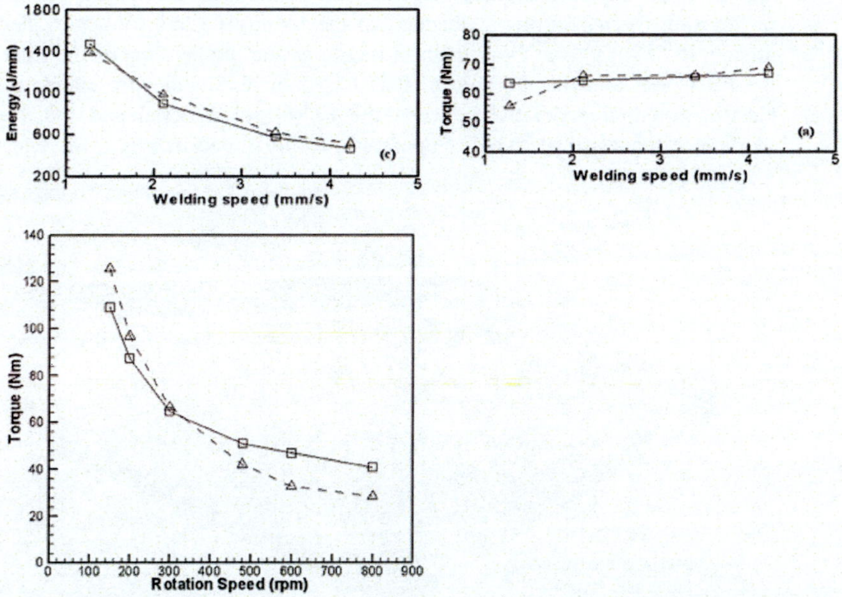

Fig. 6.16 Variation of energy and torque requirement with welding speed and rotational speed. Dashed line—experimental values reported and the solid line—numerically calculated values [8]

Research work was also done to model the FSW process which involved physical coupling between mechanics and heat transfer, large deformations and strain rates in the stirring zone around the pin. The FSW process has been modeled as first-order plus dead time (FOPDT) model, and then the model predictive control is used to analyze and elucidate the startup process, large load changes, and constrained control.

Taysom [11] evaluated FOPDT model and hybrid heat source (HHS) model using model predictive control (MPC). Model parameters were determined by fitting model predictions to actual weld data. The models were evaluated for their performance in modeled and unmodeled disturbances and also for control immediately after the plunge phase. The locations for various energy transfers are shown in Fig. 6.17. The hybrid heat source model performed better at the startup of the weld, while FOPDT was found to be of the wrong form to analyze FSW immediately after plunge. The HHS MPC model had lesser variability between two runs compared to PID regulator controller, but the PID controller held the temperature much closer to the setpoint.

Weld temperature is the key control parameter in determining the quality of the weld. Low temperatures during FSW avoid several defects typically observed in fusion welding processes such as porosities and cracks and yields good mechanical properties over arc welding. The reduced heat input also results in lower deformation. Temperature is to be considered for the feedback control designed for FSW.

Silva et al. [12] presented an overview of temperature measurement methods applied to FSW process. Three methods were evaluated: thermocouples embedded in the tool, thermocouples embedded in the workpiece and tool–workpiece thermocouple (TWT). The TWT method measures the temperature at the interface of the FSW tool and the workpiece. It is based on thermoelectric effect where the electric potential generated between the FSW tool material and the aluminum workpiece relates to the weld temperature, as illustrated in Fig. 6.18. The voltage

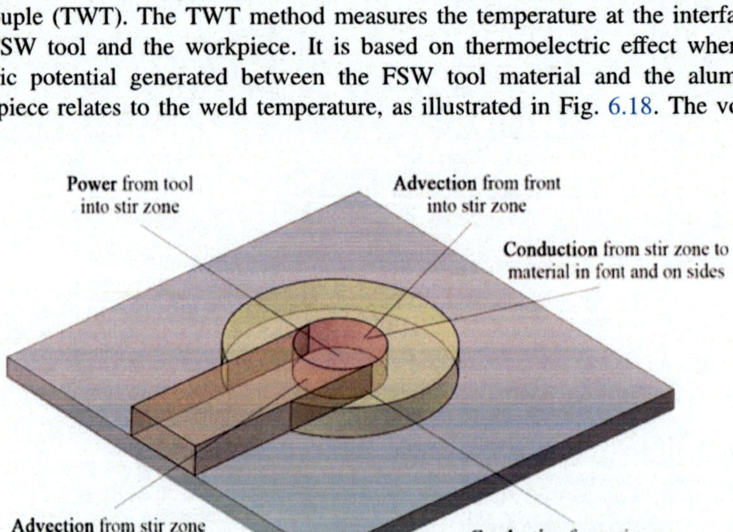

Fig. 6.17 FOPDT model [11]

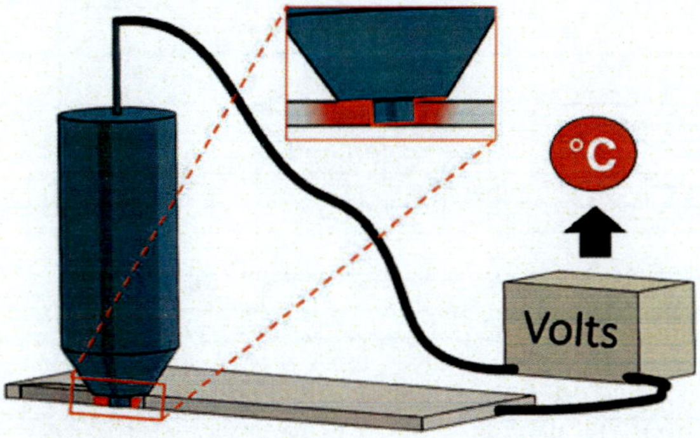

Fig. 6.18 Tool–workpiece thermocouple method [12]

measured will depend on the tool and workpiece material properties. TWT was found to be an accurate and fast method suitable for feedback control of FSW.

Mishra et al. [13] reviewed various techniques and methodologies applied for sensor-based monitoring and control of FSW process. They discussed the applicability of various sensors such as force, torque, current, power temperature, vibration, acoustic emission, and imaging to acquire information about the process. They also proposed a roadmap for implementing the idea of Industry 4.0 to the FSW process and presented a typical architecture for monitoring the parameters shown in Fig. 6.19. The weld quality is mainly affected by joining parameters, design parameters, and material parameters. The joining parameters can be

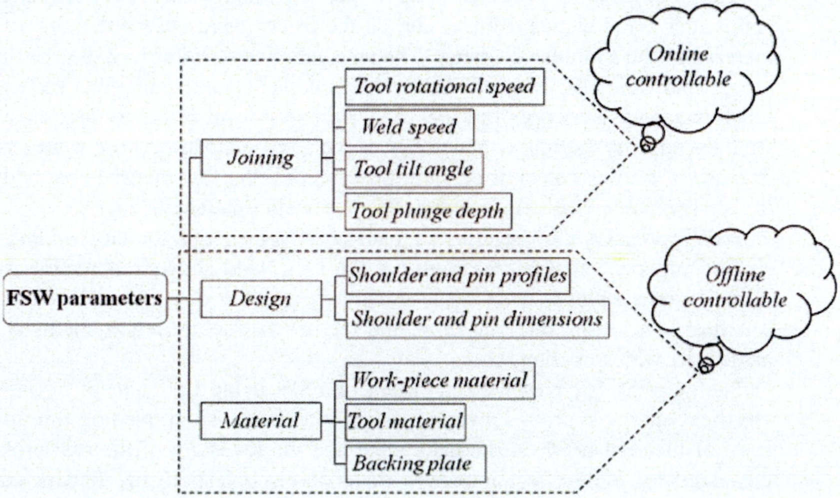

Fig. 6.19 A typical monitoring architecture [13]

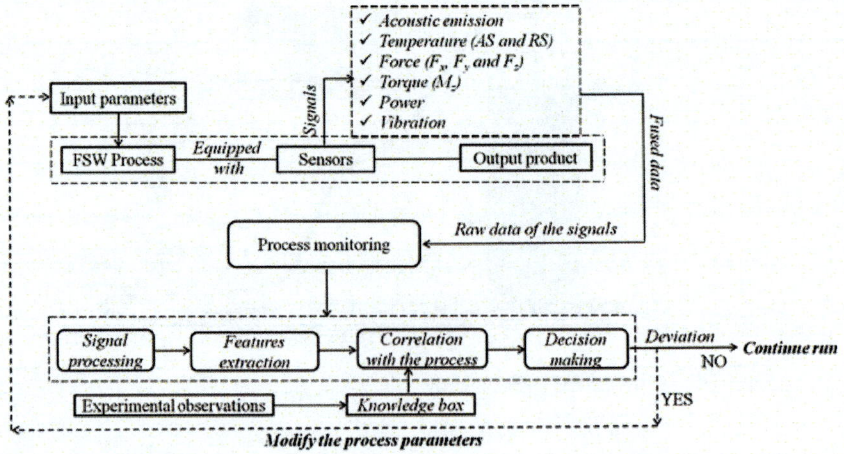

Fig. 6.20 Offline control mechanism for design and material parameters [13]

controlled in real time, while the design and material parameters have to be controlled offline as shown in Fig. 6.20.

Melendez et al. [14] developed a model to study the various forces acting in FSW. In this study, the materials having different temperature and yield strength were considered to analyze the behavior of the process forces. The different shapes of the recrystallized zones for different alloys cause positive or negative slope of longitudinal force with the plunge depth. Increase in weld speed causes increase in all the three forces. Effect of tool geometries on the various forces was found. Fluted shoulder tools were observed to increase longitudinal and transverse forces and reduce the downward force, and all the forces were observed to increase with increase in the shoulder diameter. The researchers also compared the specific weld energy and weld efficiency of FSW and fusion welds concluding that FSW requires much lesser weld energy. The specific weld energy was found to decrease with the weld speed. The thermal conductivity of the anvil plate affects the weld efficiency. The use of stainless steel or ceramic materials having low thermal conductivity for the anvil plate would help in increasing the weld efficiency.

FSW processes can be assisted with auxiliary energy for the welding of dissimilar materials. It was observed that the final weld shows remarkable improvement if preheating/presoftening is integrated into the conventional FSW process. The heat generation, material flow, and the microstructure requirements were also as desired with this integration.

Santos et al. [6] presented a variant of FSW using the concept of an external electrical energy source, delivering a high-intensity current, passing through a thin layer of material between the back plate and the lower tip of the tool probe. Heat generated by joule effect improves material viscoplasticity in this region, minimizing the root defects. The potential use of this variant was shown by reducing the

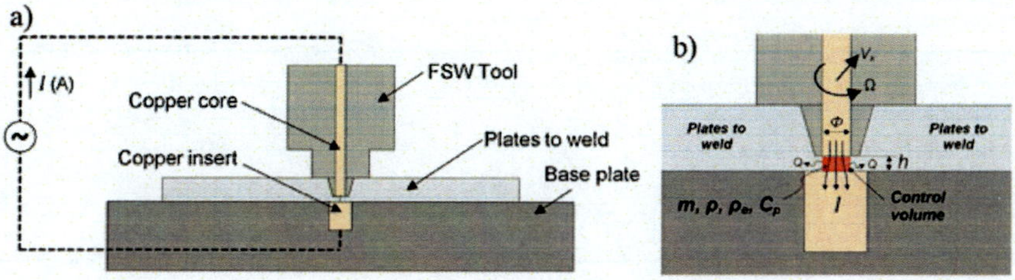

Fig. 6.21 Transverse view and detail of the parameters [6]

size of the weld root defect, even for significant levels of lack of penetration, without affecting overall metallurgical characteristics of the welded joints. Schematic of the proposed idea is shown in Fig. 6.21.

Their experimental results showed that passing an electrical current through the weld root, lack of penetration (LoP) defects reduced in size from a width of 15.5–3.3 m, under the conditions tested.

Yang et al. [19] established the temperature field model by COMSOL multi-physics to analyze the coupling effect of joule heat and friction heat on the temperature field distribution. The results showed that the temperature distribution area in the shoulder enlarges with joule heating. The maximum temperature produced by coupling function of joule heat and friction heat is higher than conventional FSW Fig. 6.22 under constant welding velocity, axial force, rotation velocity, and displacement. The joule effect-assisted FSW thus reduces the demand on spindle rigidity of welding machine and also improves the welding quality and efficiency which also reduces the energy consumption.

Parametrical analysis is imperative to analyze the dependence of the generated weld energy on the process parameters and material properties. Critical energy of the material is a significant property in this regard. The weld energy should be greater than or equal to the material critical energy. Optimal weld speed and the choice of parametric ranges are decided by this energy balance condition.

6.7 Sample Design Data, Process Parameters & Design Calculations

In the FSW process, heat generation is by direct and indirect sources. Friction heating at the interface is by direct source and the localized plastic shear deformation which happens around the tool pin is instituted by the indirect source. This forms the energy required for the welding process. This section will give a description of some FSW designs for different materials.

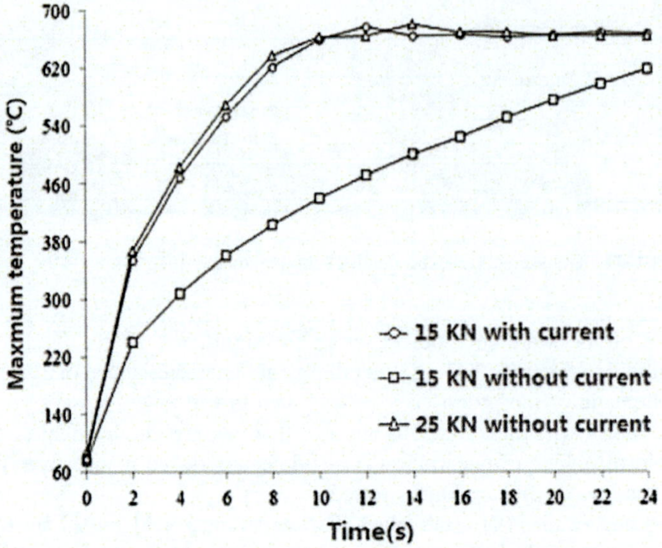

Fig. 6.22 Comparison curve of temperature with and without assisted heating [19]

6.7.1 Sample Data I

Friction stir welds were made under different weld conditions by varying plunge depth, tool rotational speed, welding speed, tool geometry, and lead angle of the tool [14].

Under typical welding conditions for 0.25 in (6.4 mm) thick 6061-T6 and 2195-T6 aluminum alloys, the values for various parameters and forces are as mentioned in this Sect. 6.2 195-T6 has been found to be difficult to be welded by fusion techniques.

Yield strength at 450 °C = 10.3 Mpa
Cross-sectional area = 0.035 in^2
Rotating tool dimensions:

- shoulder diameter—19.2 mm
- pin diameter—6.35 mm
- pin length—5.83 mm

Pin tool:

- lead angle = 1°
- Torque is given by Eq. 6.13

$$\text{Torque} = \frac{P}{2 \times \pi \times f} \tag{6.13}$$

- *P*—input power to the welder measured with a power meter
- *f*—rotational frequency

- Rotational velocity—800 rpm
- Welding speed—2 mm/s.

A digital torque wrench was used to apply torques to studs located at A and B. Strain readings were recorded with the gauges at the points A and B. A ring force gauge was used to measure the applied forces. The response of the test fixture to the known applied loads is given by Eqs. 6.14–6.16:

$$F_d = 19.3 * 10^6 \varepsilon_{31}^{Fd} lbs \tag{6.14}$$

$$F_l = 10 * 10^6 \varepsilon_{12}^{Ft} lbs \tag{6.15}$$

$$F_t = 3 * 10^6 \varepsilon_{22}^{Ft} lbs \tag{6.16}$$

The experimentally measured strains can thus be converted to determine the effective downward, longitudinal, and transverses forces. Effects of plunge depth, tool geometry, and the weld speed on the various forces are listed below are shown in Fig. 6.23 [14] and listed.

- Longitudinal Force:
  - Increases when tool rotational speed and/or weld speed is increased
  - Increases with shoulder diameter
  - 600 N for plunge depth of 0.89 mm
  - 1.5 KN for plunge depth of 1.27 mm
- Downward force—8.9KN
  - Decreases when rotational speed is increased
  - Increases with shoulder diameter
  - Decreases with welding speed for Al6061, increased for Al2195
- Transverse Force
  - Much smaller than longitudinal and downward force
  - Shows slight increase with weld speed
  - Increases with shoulder diameter.
- Readings of power versus weld speed were used to measure the specific weld energy as shown in Fig. 6.24 [14].

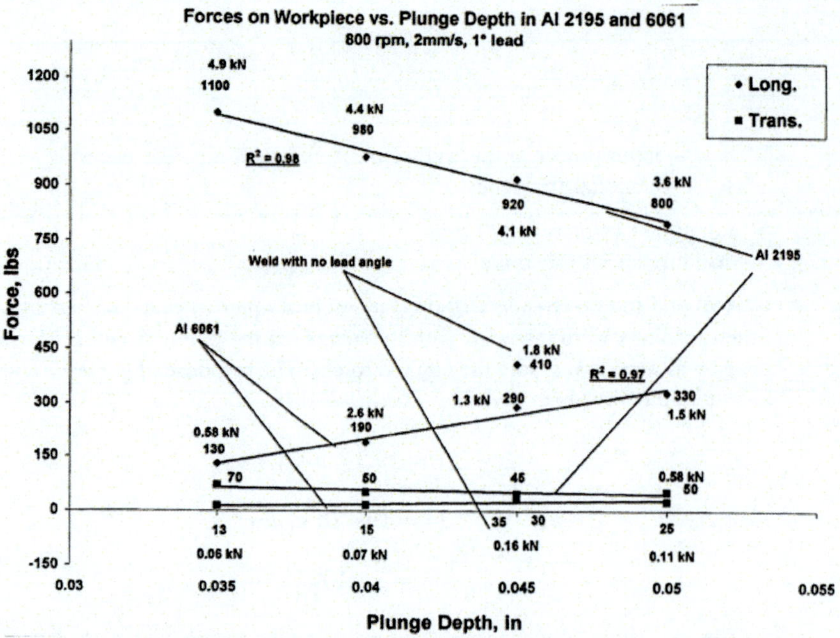

Fig. 6.23 Effect of process parameters on various forces [14]

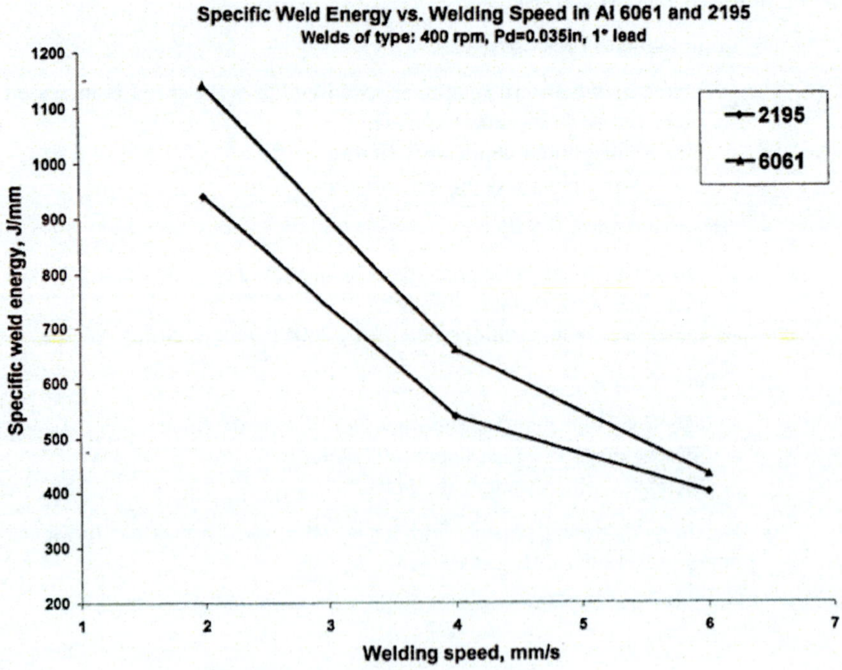

Fig. 6.24 Specific weld energy versus weld speed curve [14]

With increasing weld speed, the energy requirement per unit distance of the weld decreases. For this experiment using 0.25 inch Al 6061 at the weld speed of 4 mm/sec, energy requirement is determined to be nearly 670 J/mm.

6.7.2 Design Calculations- Case Study

The heat required for FSW process is obtained from the two sources that are friction and the plastic deformation along the weld line. In this work, the heat energy from the two sources is determined based on the process parameters and material properties. The heat generated from the two sources should be more than critical energy of the material which depends on the material properties and weld size. Based on the critical energy required, the weld parameters are designed.

The calculations shown in this section can be further extrapolated to be applied to some of the automotive applications like vapor tight fuel tank which also serves to increases the rigidity of the chassis, suspension links or foldable rear seats, to name a few.

The material properties are mentioned in Table 6.3:

Aluminum alloy Al2024-T3 [15]
Sheet thickness—10 mm

Parameters:
Speed N = 800 rpm
Pin diameter D_p = 2 mm
Shoulder diameter $D_s = 2R_p$
Contact pressure = yield strength = 325 MPa
Coefficient of friction = 0.3
Velocity = 3 mm/s

Power input due to friction between tool shoulder and the material at the interface, as depicted in the tool geometry (Fig. 6.25) is given by Eqs. 6.17 and 6.18 [18]:

$$J_f = 2\pi^2 P\alpha N\left(R_s^2 - R^2\right)\left(R_s + R\right)\frac{Nm}{s}$$
$$J_f = 2\pi^2 P\alpha NR^3\left(\phi^4 - 1\right) \tag{6.17}$$

Table 6.3 Mechanical properties of Al2014-T3

Yield strength (MPA)	Tensile strength (MPA)	Thermal conductivity (w/m °C)	Specific heat (J/kg °C)	Thermal diffusivity (m^2/s)	Density (g/cm^3)	Activation energy (J/g)
325	470	198	900	8.23e-05	2.7	5260

Fig. 6.25 Tool geometry [18]

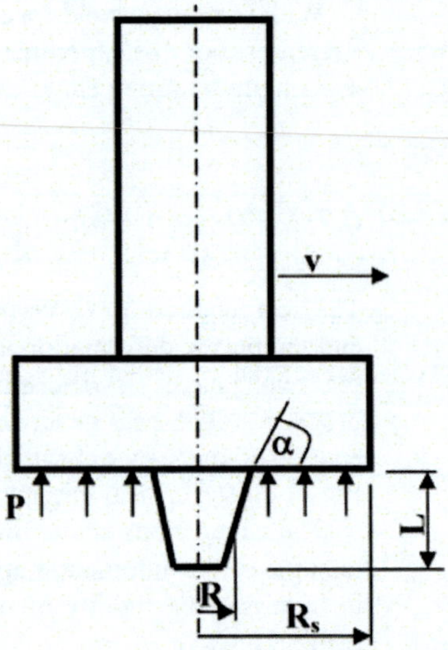

Substitute $v = 2\pi NR$ in Eq. (6.17):

$$J_f = \pi P \alpha v R^2 \left(\phi^4 - 1\right) W \tag{6.18}$$

where

alpha = coefficient of friction
P—contact pressure
N—weld speed
L—Pin length
R—pin radius
Rs—shoulder radius
Pin radius ratio $\phi = \frac{R_s}{R}$

The friction power is thus seen to increase with increase in the contact pressure (*P*), coefficient of friction (μ), pin radius (*R*), radius ratio (Φ), rotating speed (*N*) and welding speed.

For welding speed 3 mm/s, pin radius of 2 mm, shoulder radius two times of the pin radius, contact pressure equal to yield strength of the material and coefficient of friction 0.3, the friction power input can be found to be 1 kW using the curve (Fig. 6.26):

The second source of heat is the plastic deformation along the weld line caused by the pin due to its rotation and indentation. The tool pin travels along the weld with a

- translation speed V_0
- depth in the material is *L*.

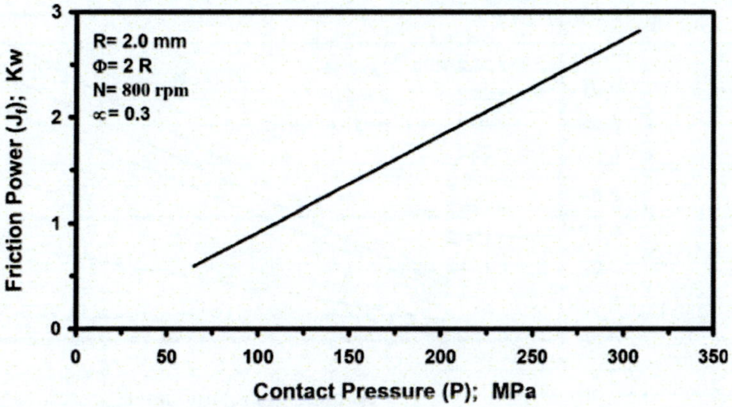

Fig. 6.26 Friction power versus contact pressure [18]

The pin causes very high plastic deformation due to its rotation and indention in the material.

The power necessary for the plastic deformation is given by Eq. 6.19:

$$J_p = \sigma€v_0(2RL) \tag{6.19}$$

Plastic deformation power depends on the following parameters:

- Flow stress of the material
- The effective strain in plastic deformation zone
- Welding speed
- Tool geometry.

Figure 6.27 shows the correlation of plastic deformation energy with the above parameters.

The experimental study could be then used to find the power input for various parameters based on Eqs. 6.1 and 6.2, and the variation is shown in the below graph (Fig. 6.28). The ratio between the friction power and the plastic deformation power for effective strain value of 18 is found to be nearly 3 given the relation: $J_f = 3J_p$

This observation establishes that the friction power is the key source of power required for welding in this process.

The total input power is given by Eq. 6.20

$$J_t = J_f + J_p \tag{6.20}$$

Power required for welding is given by Eqs. 6.21 and 6.22 for the two-dimensional and three-dimensional heat sources [20]:

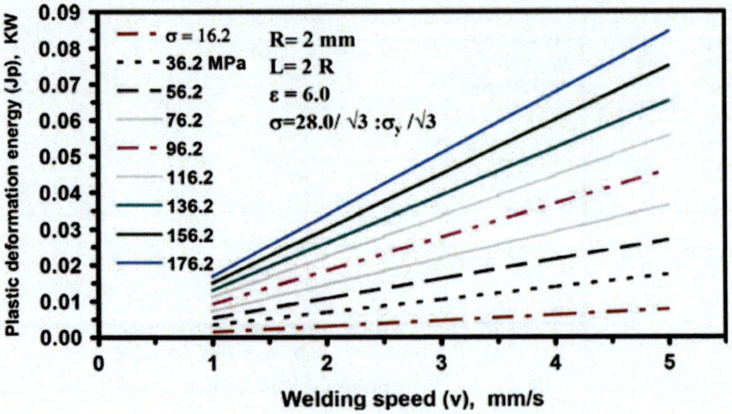

Fig. 6.27 Plastic deformation energy as a function of welding speed for different shear stress values [18]

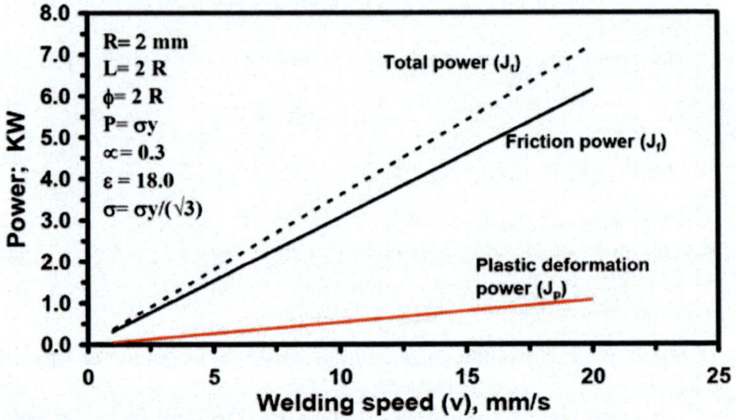

Fig. 6.28 Effect of welding speed on power Sources generated during FSW [18]

- Three-dimensional heat source

$$q = \frac{5}{4} * \pi e K T_m \left(\frac{2}{5} + \frac{vw}{4\alpha} \right) \tag{6.21}$$

- Two-dimensional heat source

$$q = 8 K T_m h \left(\frac{1}{5} + \frac{vw}{4\alpha} \right) \tag{6.22}$$

Based on the minimum power requirement criteria, the optimal values of process parameters can be determined. The energy balance condition for the required rise in temperature to cause the weld and the total heat generated is given by Eq. 6.23:

$$\dot{m}C\Delta T = J_f + J_p \tag{6.23}$$

Temperature rise can be then represented by Eq. 6.24

$$\Delta T = \frac{J_f + J_p}{\dot{m}C} \tag{6.24}$$

C—Activation energy of the material. It is the minimum energy which is responsible for diffusion of atoms and is essential for the weld to occur.

For a given material, activation energy will be known. Thus, the required temperature rise can be estimated using the material flow rate and specific heat values.

Material flow rate is an indirect measure of molecular weight and is given by Eq. 6.25

$$\dot{m} = 2RLv\rho \tag{6.25}$$

The results shown in Figs. 6.29, 6.30, and 6.31 for different width of weld and the various process parameters are used to calculate the power required for welding.

Power input due to plastic deformation is considered to be 1/3rd of the power due to friction, given by Eq. 6.26.

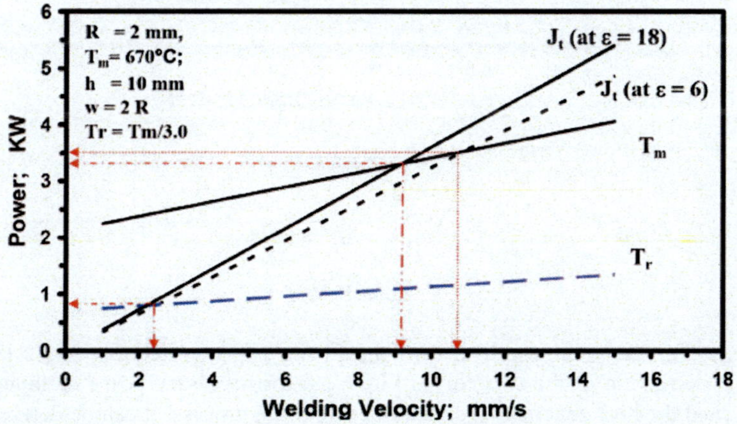

Fig. 6.29 Comparison of power versus weld speed at melting temperature and recrystallisation temperature [18]

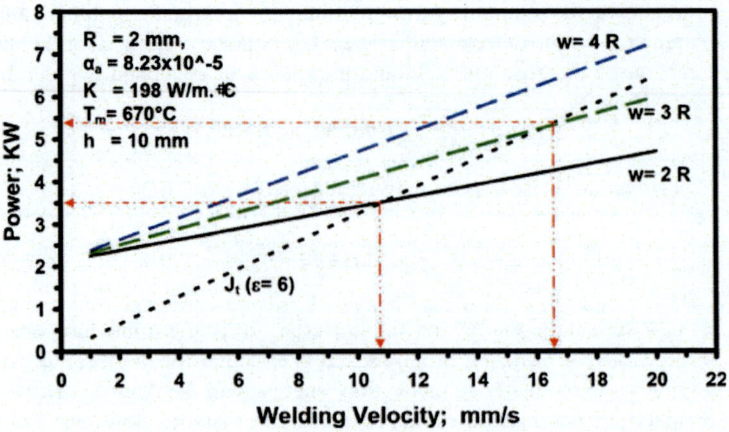

Fig. 6.30 Variation of power with weld speed at Tm for different weld widths [18]

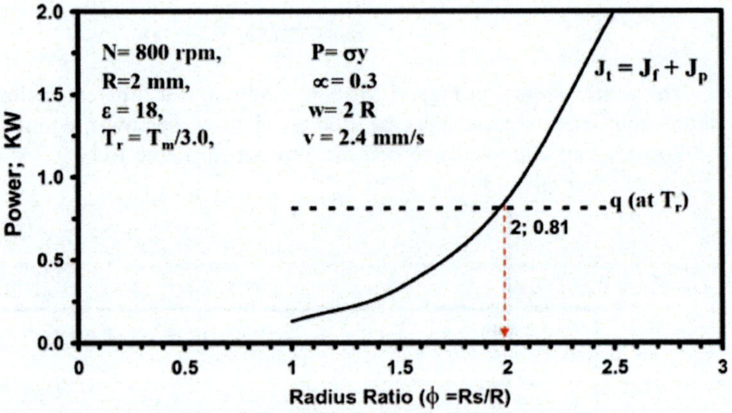

Fig. 6.31 Power versus radius ratio [18]

$$J_p = J_f/3 \tag{6.26}$$

$$\text{Total Power} = J_f + J_p$$

For the weld width of 10 mm, the input power required = 3.5 kW from the intersection of the two curves Fig. 6.29, for necessary heat rate input (solid line) and the heat generation (dotted line) from the process, it can be derived that for an Al plate of thickness 10 mm at an effective strain of 6, optimal welding speed = 11 mm/s.

To find the optimum value of process parameters, the input power should be equal to the required power shown in Fig. 6.30. Same is given by equation Eq. 6.22.

For this case, the input power is 3.2 kW.

The intersection of the two lines (J_t for ϵ = 18 and Tm) shown in Fig. 6.29 gives the optimal welding speed which should be used. For the power requirement of 3.2 kW, the welding velocity is 9 mm/sec.

Compared to the weld speeds used in previous research works, the obtained speed is three times the conventional weld speed in FSW. This high value of welding speed may be attributed to two factors:

- High prediction of heat rate input from Eq. 6.22, represented by solid line in Fig. 6.29
- High prediction of power generated from Eq. 6.20, represented by dotted line in Fig. 6.29.

The high prediction of input power and the generated power is because the required weld temperature is taken as the melting temperature. In the FSW process, since the melting of material is not allowed, thus adjustments can be made to predict lower values of heat rate input by using recrystallization temperature which is one-third of the melting temperature.

This adjustment gives predicted value of welding speed as 2.5 mm/sec (Fig. 6.31), instead of the earlier value of 11 mm/s.

Similarly, the other process parameters can be selected using this model. For example, to find the optimum radius ratio, the below characteristics shown in Fig. 6.31 can be used. The optimal value of the pin radius ratio Φ can be obtained by the point of intersection of the total energy requirement curve and the input heat rate line.

The input power required for FSW can thus be evaluated using this model and then can be compared with the actual power requirement based on the activation energy of the materials.

Activation energy of Al Q' = 32200 cal/mol = 1192.6 kcal/g = 4993.385 J/g

For a specific material flow rate, power required for activation will be given by Eq. 6.27:

$$Qa = \text{activation energy } (Q') * \text{material flow rate} \tag{6.27}$$

From Eq. 6.25,

Material flow rate $\dot{m} = 2RLv\rho = 2 * 5 * 10 * 2 * 0.0027 = 0.54g/s$

Qa = 4993.385 × 0.54 = 2696.4 W = 2.964 kW

This value is very close to the power calculated using the model [18] for welding.

6.8 Conspectus of Design Studies in Friction Stir Welding

Good design practices in welding entail knowledge of material and process requirements for accomplishing impeccable welds. This chapter is framed with an objective to impart learning and assimilating the basic design principles of another advanced joining process "FSW" followed in industrial applications. A systematic flow is adopted to understand the FSW process phenomena, process parameters and their implications on various materials, design procedures, and optimization techniques. Latest approaches to design based on algorithmic decision-making are presented that harmonizes design knowledge and experience of all relevant specialties through case studies. The following chapter is an attempt to give a comprehensive overview of the magnetic pulse welding—a solid-state welding process and covers the concepts and the design aspects of this welding technique.

References

1. Luo H, Wu T, Wang P, Zhao F, Wang H, Li Y (2019) Numerical simulation of material flow and analysis of welding characteristics in friction stir welding process. Metals 9(6):621
2. Mishra A (2018) Friction stir welding of dissimilar metal: a review. Available at SSRN 3104223
3. Liu FC, Hovanski Y, Miles MP, Sorensen CD, Nelson TW (2018) A review of friction stir welding of steels: tool, material flow, microstructure, and properties. J Mater Sci Technol 34 (1):39–57
4. He X, Gu F, Ball A (2014) A review of numerical analysis of friction stir welding. Prog Mater Sci 65:1–66
5. Helmholtz-Zentrum Geesthacht, Solid state joining processes. https://www.hzg.de/institutes_platforms/materials_research/materials_mechanics/solid_state_joining_processes/techniques/index.php.en
6. Santos TG, Miranda RM, Vilaça P (2014) Friction stir welding assisted by electrical joule effect. J Mater Process Technol 214(10):2127–2133
7. Colligan KJ (2010) Solid state joining: fundamentals of friction stir welding. In: failure mechanisms of advanced welding processes. Woodhead Publishing, pp 137–163. https://doi.org/10.1533/9781845699765.137
8. Arora A, Nandan R, Reynolds AP, DebRoy T (2009) Torque, power requirement and stir zone geometry in friction stir welding through modeling and experiments. Scripta Mater 60 (1):13–16
9. Ross KA (2012) Investigation and implementation of a robust temperature control algorithm for friction stir welding
10. Schmidt HNB, Hattel J (2004) Heat source models in simulation of heat flow in friction stir welding. Int J Offshore Polar Eng, 14(04)
11. Taysom BS (2015) Temperature control in friction stir welding using model predictive control
12. Silva ACF, De Backer J, Bolmsjö G (2017) Temperature measurements during friction stir welding. Int J Adv Manuf Technol 88(9–12):2899–2908
13. Mishra D, Roy RB, Dutta S, Pal SK, Chakravarty D (2018) A review on sensor based monitoring and control of friction stir welding process and a roadmap to Industry 4.0. J Manuf Process 36:373–397

14. Melendez M, Tang W, Schmidt C, McClure JC, Nunes AC, Murr LE (2003) Tool forces developed during friction stir welding
15. Hahn M, Weddeling C, Lueg-Althoff J, Tekkaya AE (2016) Analytical approach for magnetic pulse welding of sheet connections. J Mater Process Technol 230:131–142
16. Khaled T (2005) An outsider looks at friction stir welding. Fed Aviat Admin 25:27–29
17. Essa ARS, Ahmed MMZ, Mohamed AKYA, El-Nikhaily AE (2016) An analytical model of heat generation for eccentric cylindrical pin in friction stir welding. J Mater Res Technol 5 (3):234–240
18. El-Domiaty A, Abd El-hafez H (2015) An energy model for friction stir welding
19. Yang J, Dong L, Tian XC (2014) The effect of joule heat on coupling temperature field of friction stir welding. In: Advanced materials research, vol 941. Trans Tech Publications, pp 2043–2046
20. Mechanica Technical Solutions. Solid state welding. https://www.mechanicatech.com/Joining/solidstatewelding.html. Viewed on 23 July 19
21. Bradley GR, James MN (2000) Geometry and microstructure of metal inert gas and friction stir welded aluminium alloy 5383-H321. University of Plymouth, pp 1–78
22. Thomas W, Russell MJ, Duncan A, Robelou A, Park G (2010) Friction stir welding, an introduction to innovative variant techniques for the aluminium industry. In: International aluminium congress and exposition. Queretaro, Mexico
23. Thomas WM, Johnson KI, Wiesner CS (2003) Friction stir welding–recent developments in tool and process technologies. Adv Eng Mater 5(7):485–490
24. Lakshmi AA, Rao C, Kotkunde N, Subbiah R, Singh SK (2018) Forming limit diagram of AISI 304 austenitic stainless steel at elevated temperature: experimentation and modelling. Int J Mech Eng Technol (IJMET) 9(12):403–407

Chapter 7
Magnetic Pulse Welding and Design

Abstract Advent of a polyvalent solid-state cold-welding process titled "Magnetic Pulse Welding (MPW)" has un-wrapped scope for reconnoitering uncommon metallurgical behaviors. For a researcher investigating MPW welding with a view to applying it for non-ferrous and dissimilar materials, the available literatures is to some extent, sporadic, and scanty. Many research articles are translated from other foreign languages. In several cases, the translations are not clear; accumulating to the difficulty in understanding. This chapter attempts to proffer a comprehensive overview of the earlier research work carried out in the area of MPW welding. Besides, it divulges the disagreement among various researchers with respect to the underlying concepts of MPW welding; in particular, the discrepancies pertain to the interaction between the flyer, the target, and induced magnetic fields. The chapter commences with the initial postulates on MPW welding process and reports the design/developmental aspects. Further, the applications and research advances in MPW are covered. The design approaches with sample calculations of the weld process are included to determine the optimum process parameter ranges for various configurations of MPW.

7.1 Introduction

Magnetic pulse welding (MPW) is a cold-welding process which relies on the electromagnetically generated high velocity impact to join two metal parts. The electromagnetic force accelerates one specimen (flyer) toward the other weld specimen (target) creating the requisite impact and subsequent metallic bond between them.

The specimens undergo transformation to semi-viscous state and eventually penetrate into one another. In the absence of melting of the workpieces, no fusion occurs. In corollary, the thermal changes influencing the material properties are negligible in this process. This distinctive feature of electromagnetic welding facilitates the joining of dissimilar materials with different melting points while retaining their properties.

M. Chaturvedi and S. Arungalai Vendan, *Advanced Welding Techniques*,
https://doi.org/10.1007/978-981-33-6621-3_7

A capacitor bank is charged up to the energy required for joining the prescribed material combination. It acts as a pulse generator and provides high-frequency and high intensity alternating current.

Figure 7.1a shows a schematic diagram of a current discharge circuit [1]. The circuit consists of a capacitor bank for supply of electrical energy, a discharge gap switch, and an *E*-shaped one-turn flat coil. The two plates are placed above the coil, in proximity.

The two plates are named as flyer plate (movable plate) near to the coil and the target plate (fixed plate). With the capacitor fully charged, and the discharge gap switch closed, a discharge pulse is released to the coil. The discharge energy stored in the capacitor is the key parameter governing the welding process. Figure 7.1b illustrates the discharge pulse, eddy current, magnetic flux, and the emf.

When a discharge pulse from the capacitor passes through the coil, the change in the flux linkage causes a high-density magnetic field to be created around the coil. The magnetic flux lines intersect with the flyer plate, and induce an electromotive force that gives rise to a current in the flyer and the magnetic field created in the flyer opposes its cause. The Lorentz force created acts in the upward direction due to the current flowing in the primary magnetic field. The MPW arrangement with the electromagnetic forces responsible cross-sectional view of the process is illustrated in Fig. 7.2. The flyer plate is subjected to this force and is pushed toward the target plate with a high velocity (approximately 300 m/s). MPW process is thus a high velocity forming process. As a consequence of the impact of the flyer plate on the target plate, the flyer undergoes plastic deformation causing solid-state weld under controlled conditions [2].

The two materials collide at a certain angle at high velocities and produce an impact pressure as shown in Fig. 7.3.

The choice of the target and flyer plates is based on the electrical conductivity. The material having lesser electrical conductivity will be used as the target.

Conventional fusion-based welding techniques have certain limitations due to the microstructural and mechanical changes in the weld bead and heat affected zones resulting in reduction of joint strength and also formation of hot cracks [2, 3]. MPW is advantageous owing to the formation of metallic bonding of chemically pure substances with proper impacting parameters. There is no requirement of a

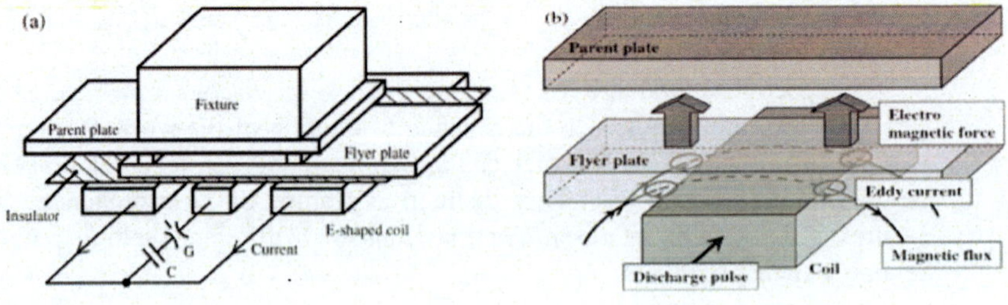

Fig. 7.1 **a** Schematic of current discharge circuit, **b** middle section close up [2]

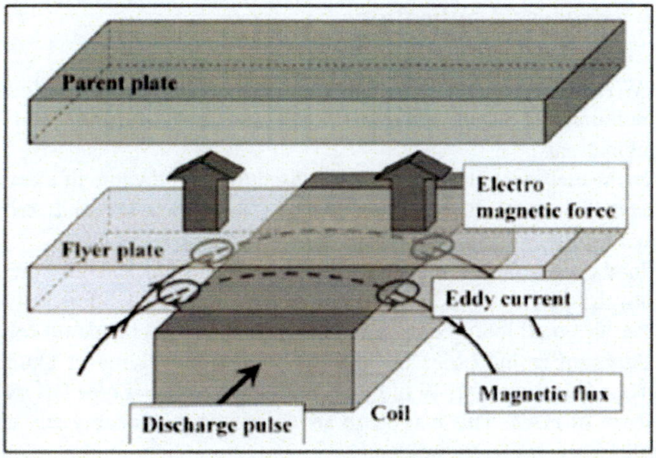

Fig. 7.2 MPW arrangement showing electromagnetic effects [8]

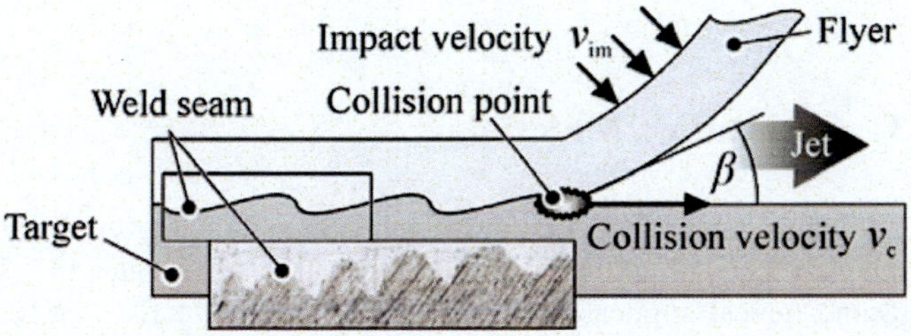

Fig. 7.3 MPW schematic [3]

filler material or a tool traversing the workpiece for implementing the weld. The problems that arise with heat affected zones and intermetallic phases are eliminated in MPW. This process has no shielding gases and shows good reproducibility. The process can be automated to improve the efficiency and achieve better control.

However, it has the following restrictions [4]:

- The flyer plate must have a good electrical conductivity, to accomplish optimum energy requirement and process cost.
- The inner material must have sufficient mechanical strength to withstand the high velocity impact without undergoing deformation.
- For safe handling of the high currents and voltages, appropriate electrical safety measures are to be espoused.

7.2 Process Applications

MPW does not involve heat, thus may be used for joining dissimilar materials like aluminum/steel and aluminum/copper and ferrous/non-ferrous material combinations.

In the electrical industry, MPW is used for manufacture of electrical fuses, motor components, cable ducts, connectors to copper cables, and termination joins of coaxial cables [4, 5].

In the automotive industry, MPW is found suitable for the development of lightweight vehicle bodies and components like fuel filters, tubular seat components, air conditioning components, and reinforcing bands on oil filters.

Automotive industries use these systems for welding of aluminum for HVAC parts or fuel filters. MPW has an advantage over other types of welding because of absence of HAZ. This can be an advantage when probably any component with a plastic filter has to be welded.

This technique has the potential to replace some of the existing technologies like brazing, friction welding, roll bonding, and explosive welding [4]. Advancements in this technology are in progress for its utility in the manufacturing of composites.

7.3 Compatible Materials

MPW yields good results in terms of strength and quality for aluminum and copper alloys that have a high thermal and electrical conductivities [5, 6].

Some typical applications may include aluminum EN AW1050 as flyers and copper CU-DHP, stainless steel 1.4307 as targets.

Multi-material welds as aluminum-iron, nickel and copper, aluminum and steel or magnesium, aluminum and titanium can be obtained with this process [6].

Skin depth, the important term in MPW, is defined as the depth up to which the magnetic field penetrates inside the flyer metal from the top of its surface. The magnetic field is observed to be concentrated more at the surface of the conductor (flyer) and decays exponentially as it moves toward the other edge.

7.4 Fundamentals of Magnetic Pulse Weld Process

In MPW, the atoms of the involved materials are impacted to such an extent that they exchange valence electrons. As a result, a wavy interface morphology is observed [3].

The discharge current from the capacitor bank is a damped sinusoidal wave and leads to development of a time-varying magnetic pressure as shown in Fig. 7.4.

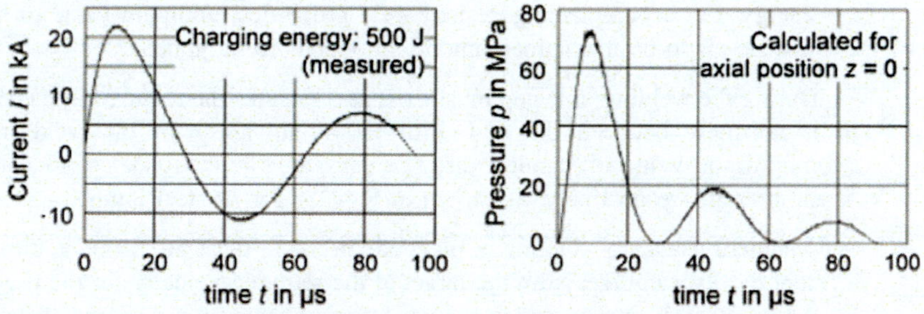

Fig. 7.4 Discharge current and magnetic pressure variation with time [3]

The magnetic pressure thus developed will also vary based upon the axial and circumferential position of a field shaper that may be used to increase the magnetic field intensity.

Suitable identification of impact parameters—angle and velocity—is determined based on the material to be welded and corresponding energy required. Systematic adjustments of impact parameters relies on the equipment design parameters and charging energy.

The main process parameters are:

- Discharge energy—This energy enables the flyer metal to move. The critical strain rate of the material should be regulated as exceeding the standard value can tear them apart, thus emphasizing on controlled energy.
- Standoff distance—This is the original distance between the two parts—flyer plate and the target plate as shown in Fig. 7.5. This space enables the flyer plate to gain velocity for the required transformation of kinetic energy to impact

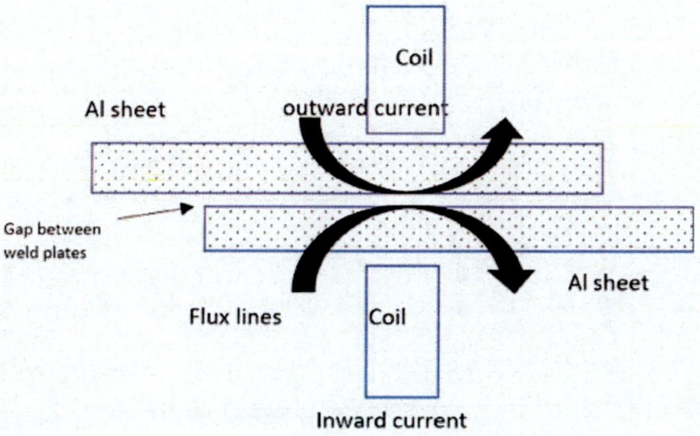

Fig. 7.5 Sample showing center of welding joint

energy. For accomplishing desired weld properties, optimum value of standoff distance is to be maintained among the metals to be joined.

During the welding, this gap must exist as it ensures the required impact energy. It is normally between 0.5 and 3 times the thickness of the weld material. Non-optimum value of standoff distance can cause reduction of weld width and shear strength, respectively, as shown in Fig. 7.6 for Al/steel joints.

- Magnetic pressure—Optimum magnetic pressure must be maintained to ensure that the flyer collides with the target at the required velocity for the bonding to occur. High discharge energy or high-frequency current can provide the required magnetic pressure. Increased magnetic pressure also increases the tensile strength of the weld.
- Impact velocity—The standoff distance, discharge energy, and time of the capacitor affect the impact velocity. For an effective weld, the impact velocity should satisfy the requirements for the complete energy transfer. Impact velocities in this process range from 200 to 700 m/s. The collision should be subsonic.
- Collision angle—The two materials should collide at certain angle and at a high velocity so as to create jetting resulting in formation of the weld.
- Skin depth is the depth through which the interaction of both the magnetic fields is limited within the workpiece (flyer). Due to this skin effect, the repulsive magnetic field produces electromagnetic forces that exert the requisite pressure on the flyer.
  - The mathematical representation of skin depth is given by Eq. (7.1)

$$\delta = \frac{1}{\sqrt{\pi \sigma \mu f}} \tag{7.1}$$

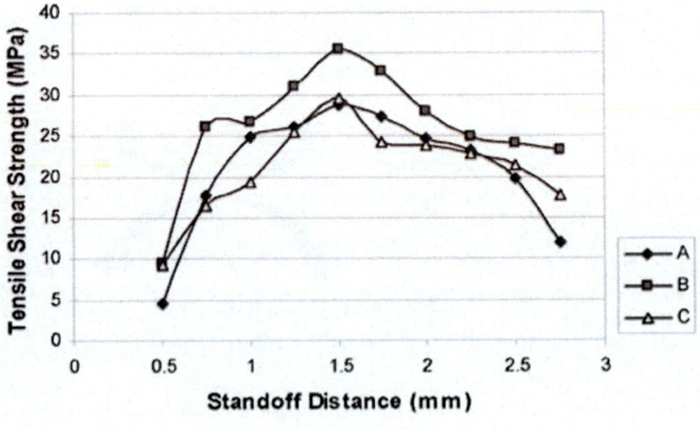

Fig. 7.6 Variation of shear strength with standoff distance [4]

where δ is the skin depth (m), σ conductivity of conductor (mho m), μ is the absolute magnetic permeability of the conductor (H/m), and f is the frequency of the current (Hz).

Figure 7.7 presents the relationship between skin depth and resistivity for several metals. At lower frequencies, skin depth is inversely proportional to material conductivity.

- Thus, it is important to employ a high frequency when low conductivity materials are employed. The skin depth should be less than the wall thickness of the workpiece. On neglecting this criteria, a part of current may not have a surface to flow in, and thus, the induced magnetic field will be less effective resulting in a poor weld [7].

7.5 Magnetic Pulse Welding Machine Details

The MPW system consists of a high-voltage power supply, a bank of capacitors, a gap switch with high switching frequency, control and trigger system, and a coil [7] as shown in the block diagram of Fig. 7.8.

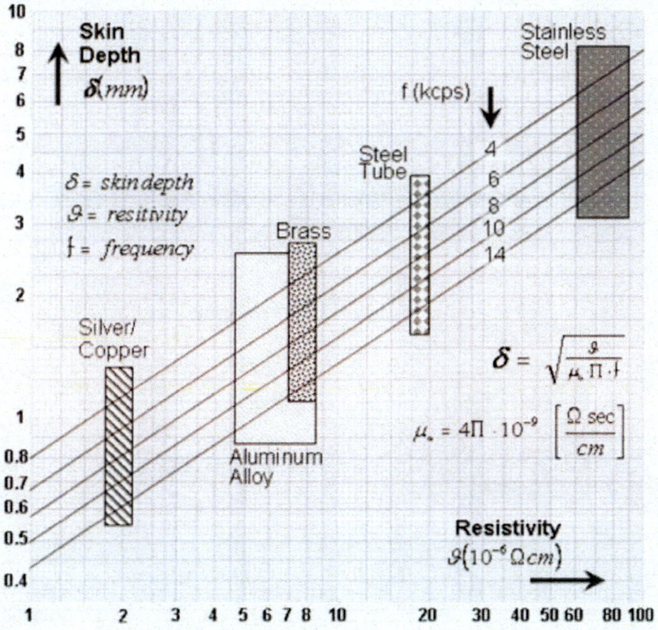

Fig. 7.7 Skin depth and resistivity relation [7]

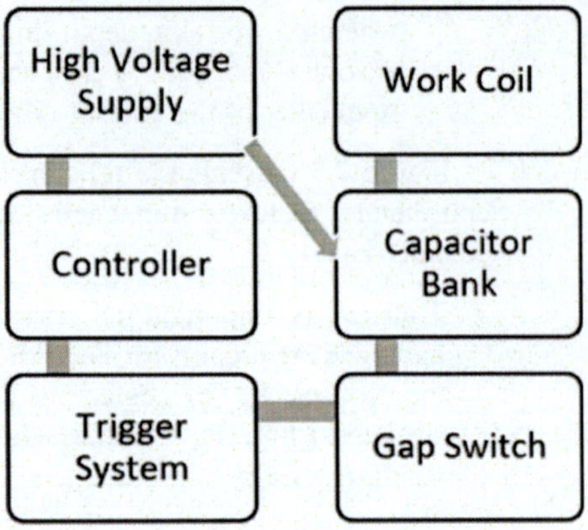

Fig. 7.8 MPW system

The complete setup can be divided into the following units as shown in Fig. 7.9:

- Control cabinet—charging unit
- Pulse generator—bank of capacitors with gap switch
- Workstation—electromagnetic coil, workpiece, electrical cables, field shaper, and transformer.

The transformer increases the voltage level to appropriately charge the capacitor bank. This energy is then discharged into welding coil. In the positive half cycle,

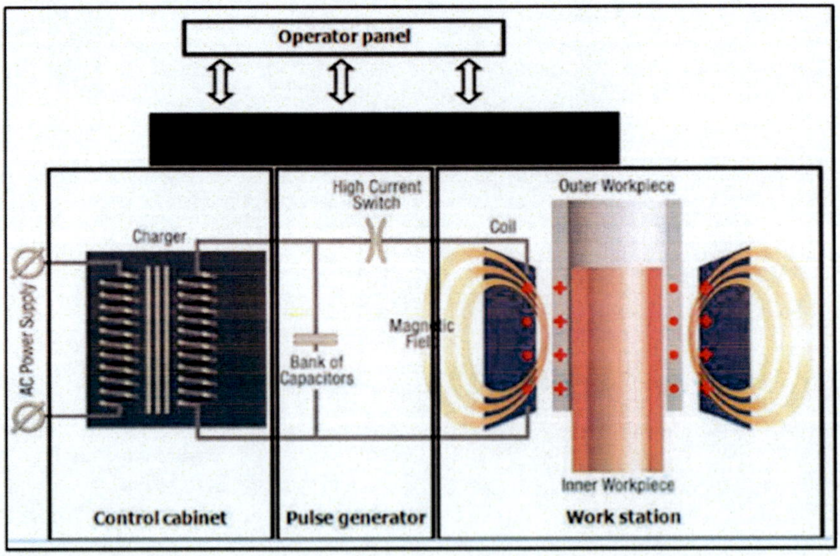

Fig. 7.9 MPW setup for tubes [7]

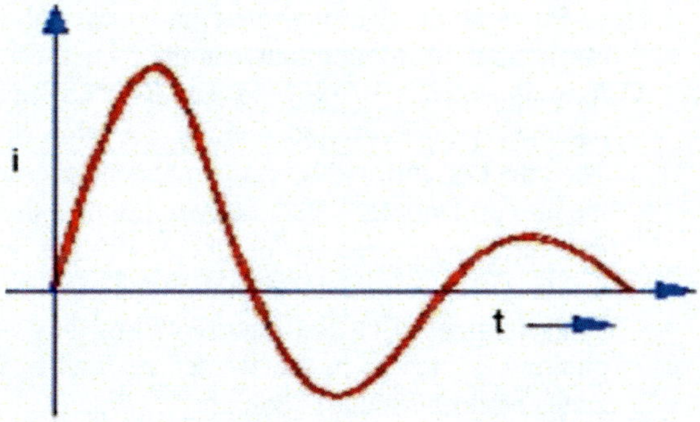

Fig. 7.10 Current waveform in an underdamped circuit [8]

the energy is delivered from the capacitor to the inductor, and in the negative half cycle, the inductor returns the energy back to capacitor. However, the current decreases continuously due to dissipation in the resistance of the circuit. The underdamped circuit current waveform looks like in Fig. 7.10.

This produces a high current in the range of few hundred kA as pulses in the coil through a control switch within 100 μs.

Electromagnetic coil is used to discharge the extreme levels of current. Design of the coil is very critical since the weld quality depends upon it. The coil design includes shape, electrical, and mechanical properties of the materials. The coil is made up of high electrically conductive material in order to generate a high intensity magnetic field [2, 8].

The coil used in the setup may be a compression coil or an expansion coil or a plate coil.

The workpiece is placed in proximity to the coil and has good magnetic coupling with the coil. According to the principle of mutual inductance of transformer, the welding coil behaves like a primary winding of the air core transformer, and the workpiece behaves as a short-circuited secondary. This causes the discharged current to pass through the working coil (in opposite direction). Induced currents in the workpiece are anti-phase with respect to the coil current.

The disadvantages usually associated with MPW are related to equipment and running costs of the coils (limited lifetime), as well as positioning systems that must be precise and specially designed for any specific applications.

7.6 Literature Survey

The MPW process has been implemented by a few manufacturers; however, it is still in preliminary stage of adoption by key industrial sectors. Many researchers have taken up MPW to explore modifications in design for technical feasibility and economic viability for applying it to various applications. Some of the prominent literature reports on MPW are briefed in this section.

Hahn et al. [3] presented an analytical model which would determine the magnetic field strength between the flyer sheet and the target sheet to determine the forming pressure in the weld process. They investigated the effects of experimental parameters on electromagnetic acceleration of 5000-series aluminum alloy sheets. The effect of coil width versus magnetic pressure is shown in Fig. 7.11. They used the photon Doppler velocimetry to confirm the validity of the model. They showed that an impact velocity of about 400 m/s is necessary for the magnetic pulse welding of 1 mm-thick EN AW 5005A sheet onto an EN AW 6060 hollow profile. With the help of etched microsections, they confirmed that a wavy interface morphology is present in the welded regions in which no interlayers, voids, or melt zones could be found.

It can be seen that the pressure theoretically increases till infinity for an infinitely large current and decreases to zero for a wider coils. The curves reveal the existence of a high-frequency limit, indicating that the current entirely flows on the coil surface near the flyer plate.

Broeckhove et al. [9] investigated an analytical model to observe the parameters that control the MPW process. They worked on Pulsar model, developed by the manufacturer of the welding machine. They arrived at a set of values for the impact velocity and collision angle and were able to identify regions in the curve for smooth or wavy interface region as shown in Fig. 7.11. The shaded area provides the operating range to obtain a successful weld.

RLC circuit represents the discharge circuit. Each element of the circuit represents the corresponding element of the welding equipment. For the coil and the workpieces, resistance was considered while inductances were considered for the coil, field shaper, and workpiece. Finally, the capacitance of the capacitor bank was accounted. Further, all the calculations for the required acceleration, pressure,

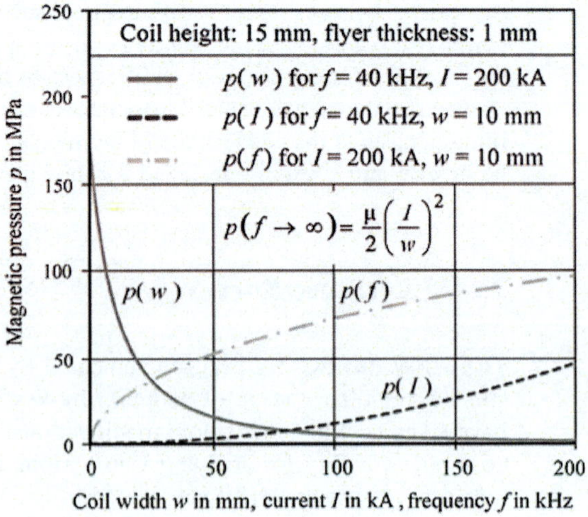

Fig. 7.11 Effect of coil width on magnetic pressure [3]

magnetic field, and voltage to be stored by capacitors were performed based on analytical concepts.

Psyk et al. [10] focused on the MPW of a tubular component to an outer joining surface by electromagnetic expansion. They used an expansion coil in their experiment and photon Doppler velocimeter for calculating the expansion velocity. Depending on the setup and the coil geometry, the technology can be applied for the compression and expansion of tubular workpieces and hollow profiles as well as for the forming of flat or three dimensionally preformed sheet metal. Figure 7.12 shows the results of the experiments carried out for different axial positions, variation of current, and radial displacement with time.

They also analyzed the dependency of the impact angle and the weld quality on the process parameters. They also worked to infer the mechanisms required for the impact process and joint design. They used the Cornerstone software to determine the effect of the influencing parameters on the impacting conditions. The results are as shown in Fig. 7.13.

Acceptable weld quality requires a minimum capacitor discharging energy based upon the material combination. Optimum value of initial gap width and flyer edge position affect the weld quality. They reinstated that the material combination is one factor for deciding the optimum standoff distance. There may even be multiple optimum values since the impact velocity and the angle both depend upon the gap. The impact velocity and angle influence the weld quality in opposite manners, and the effect of gap on the impact angle is stronger than on the impact velocity. They also identified a precise value of −2 mm. Their results show that smaller thickness of the flyer material ensures an optimum weld quality.

Kapil and Sharma [7] worked to enlist the problems and challenges that lie in the MPW process and also indicated the possible research requirements in this field. Their study gave clarity on the limitation due to the skin depth and suggested that the use of driver plates can help in creating optimum welds by providing the

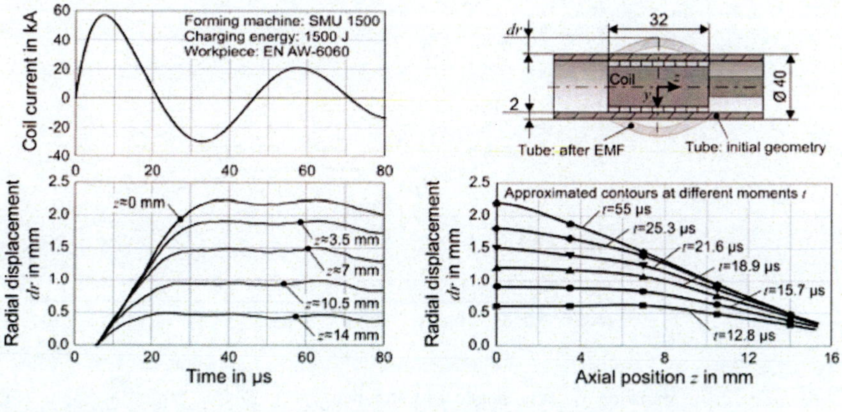

Fig. 7.12 Experimental results for various axial positions and variation of time [10]

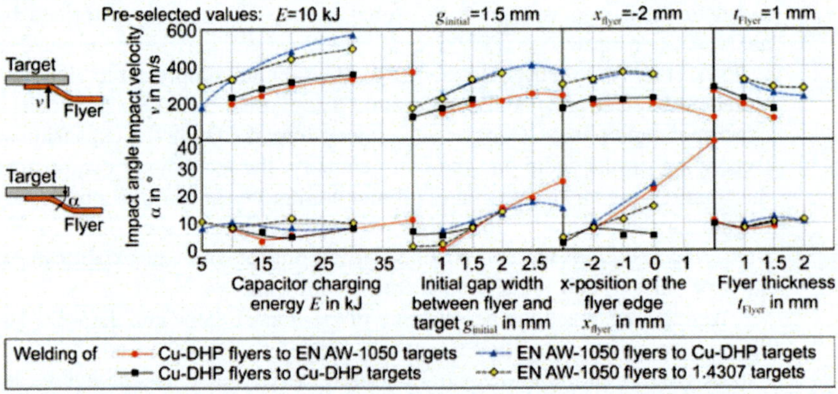

Fig. 7.13 Effect of varying influencing parameters on the impacting conditions [10]

necessary acceleration, for example, use of aluminum drivers to perform the MPW of magnesium to aluminum. The magnetic field was found to be highest at the surface of the conductor (flyer) and decays exponentially as it moves toward the other edge. As a result, the shielding of the magnetic field by the flyer metal will be poor, and sufficient radial forces will not be set up on the flyer, thus resulting in an inefficient process. Figure 7.14 shows the variation of skin depth of various materials as a function of current frequency. The compatible combination of material and current frequency should be chosen to achieve smaller skin depths. The induced magnetic field will be less effective if the skin depth is large and may result in less effective deformation.

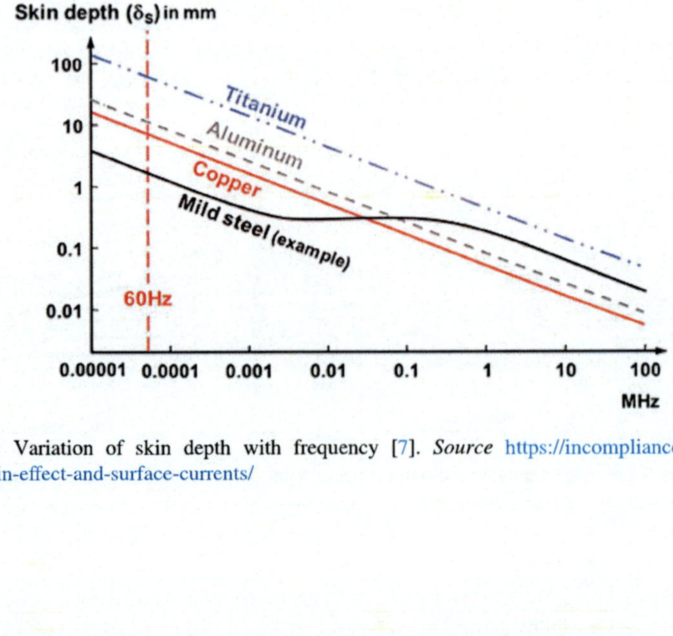

Fig. 7.14 Variation of skin depth with frequency [7]. *Source* https://incompliancemag.com/article/skin-effect-and-surface-currents/

Khanolkar [8] aimed to assess the weldability criteria of high-speed magnetic pulse welding for tubular jobs of Al, Cu, and SS combinations using finite element analysis. They proposed a circular design of EMW coil to perform EMW simulations while varying the air-gap between the outer tube and inner tube of different workpieces and voltages. Welding simulations were established between the similar metal combinations of stainless steel–stainless ateel, and between dissimilar metal combinations of aluminum–stainless Steel, and copper–stainless steel. CAD model for the working coil with tubular workpieces is as shown (Fig. 7.15)

Simulation results for Al-SS workpieces with Cu coil (Fig. 7.16): Cu Coil, AL—SS workpieces, 0.5 mm air-gap

They concluded that the values of magnetic lines of flux for different sets of combinations of coil materials for copper and stainless steel, for a given value of air-gap, are the same. Stainless steel coil shows higher resistance to the deformation during welding and thus is preferred over copper coils for the same values of voltage and air-gap. This analysis gives important inputs for the predictive design and the standardization procedures.

Thibaudeau [11] proposed an alternative design process by using an analytical model and a FEA structural analysis in an attempt to resolve the problem of single turn coil causing localized pressure distribution in a small region. They analyzed and designed an electromagnetic actuator based on the work done by Kamal and Daehn [12] for the uniform pressure actuator (UPA). This was observed to provide uniform pressure distribution and is also a robust arrangement for the welding process. This actuator has a helical coil with a rectangular cross section and a

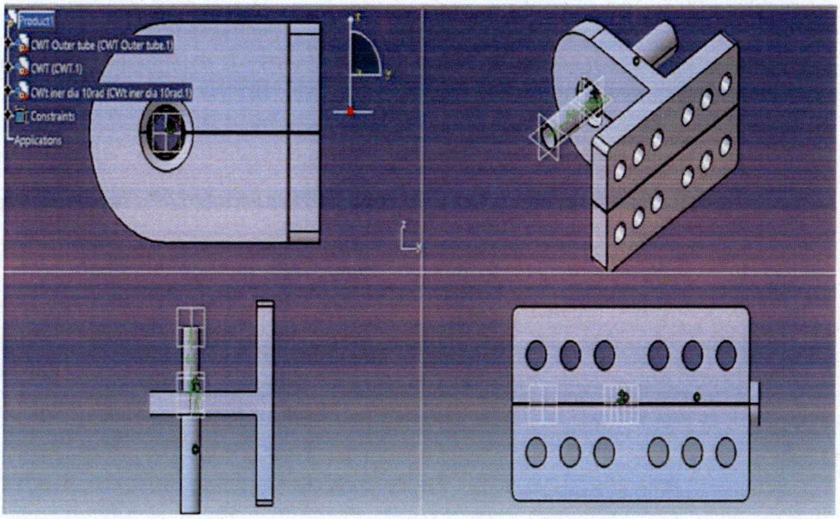

Fig. 7.15 CAD model for the working coil with tubular workpieces [8]

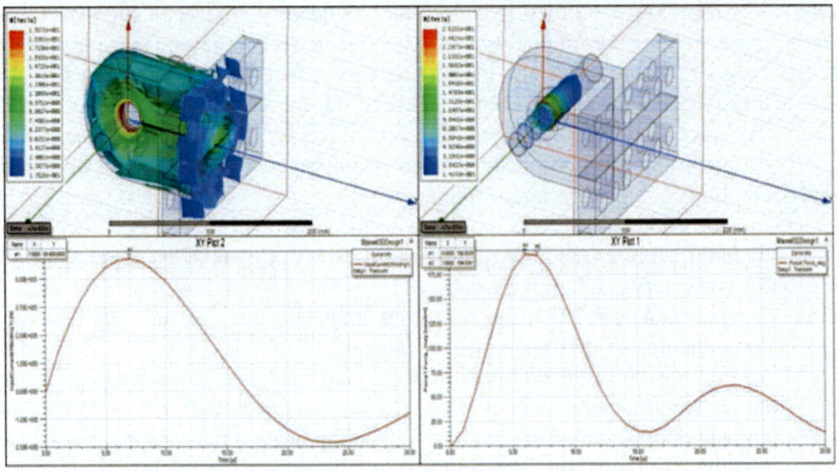

Fig. 7.16 Simulation results for Al-SS workpieces with Cu coil [8]

surrounding conductive channel to allow induced eddy currents in the sheet to form a closed circuit around the coil, thus avoiding any edge effects.

Despite the process benefits, the industrial use of MPW is restricted for applications in various industries. This could be attributed to insufficient design guidelines which extensively cover the components, materials or the process. The interdependencies of the process parameters, workpiece characteristics, and equipment behavior could be the reason for unavailability of definite guidelines for operation of this process.

This chapter presents the process feasibility and benefits. Nevertheless, investigations have to be stepped up for behavioral analysis for individual welding cases and various material combinations.

7.7 Sample Design Data, Process Parameters, and Design Calculations

The publication of analytical results provides valuable information that may be adopted for further examination by researchers. Some of the research reports which were referred to arrive at the calculations of optimum process parameters are described in this section.

7.7.1 Design Calculations I

Establishing relation between magnetic field, pressure, current, and supply voltage [13, 14]

The impact velocity and the impact angle are the significant parameters affecting the weld quality. These parameters are affected by the material geometry, its characteristics, and the electrical circuit response. In the MPW machine, supply voltage can be varied, which further affects the magnetic field, the force exerted on the flyer, and thus, the impact velocity.

The capacitor gets charged up by the supply, and the energy stored is given by Eq. (7.2).

$$E = \frac{1}{2}CV^2 \tag{7.2}$$

E—stored energy, *C*—capacitance, *V*—supply voltage.

The capacitor on charging to its maximum capacity, starts releasing energy through the discharge coil. Subsequently, a damped sinusoidal current result and the first peak of the current cause the magnetic force to be generated and thus impart impact velocity to the flyer plate. The size of the capacitor bank determines the frequency of this current based on this relation and is given in Eq. (7.3).

$$f = \frac{10^6 * I}{2\pi V * C}\ \text{Hz} \tag{7.3}$$

The peak current generated in the damped sinusoid can be estimated with the RLC equations. And thus, supply voltage is given by Eq. (7.4):

$$V = \frac{I}{\delta\sqrt{\frac{C}{L}}} \tag{7.4}$$

C—circuit capacitance, *L*—circuit inductance, δ—attenuation constant.

Attenuation constant indicates the amount by which the oscillation in a circuit gradually decreases over time.

Supply voltage level determines the impact velocity for a given material and its geometry. Optimum voltage level should be chosen to obtain a good weld.

This current in the coil induces a transient magnetic field around the workpiece. The magnetic field is assumed to be uniform between the coil and the flyer workpiece. Using Ampere's law:

Magnetic field is given by Eq. (7.5)

$$B = \mu H = \frac{\mu NI}{l} \tag{7.5}$$

B—magnetic flux density, *n*—number of coils, *I*—peak current, *l*—length of the coil, *μ*—permeability of material, *H*—magnetic field intensity.

Field shaper enables concentration of the magnetic field. Its effectiveness is determined by its inner diameter and the effective working zone. Figure protrusions, called nodules, are made on the surface of the field shaper to enable the concentration of the magnetic field on the workpiece.

Current and the magnetic field are also related by the expression 7.6:

$$B = K.I.f(A_{\text{eff}}) \tag{7.6}$$

K—circuit constant, $f(A_{\text{eff}})$—function of field shaper length unit H/m^2.

Field shaper function is found to have the variation as depicted in Fig. 7.17. This curve shows the variation of the enhancement factor of flux density as a function of nodule width or the work length.

Circuit constant *K*-would vary based upon the materials chosen for different experiments. Thus, analysis of current in the coil was done to arrive at a direct function of electrical and material characteristics. The simulation results indicated that the maximum flux density is obtained in front of the nodules of the field shaper.

If the required magnetic field is known, the above relations can be used to find the required peak current value. This current in the coil can be split up into two components—one in the working zone and other near to the ends which accounts for the energy loss in the edges. Equations (7.6–7.10) describe the equations for the currents.

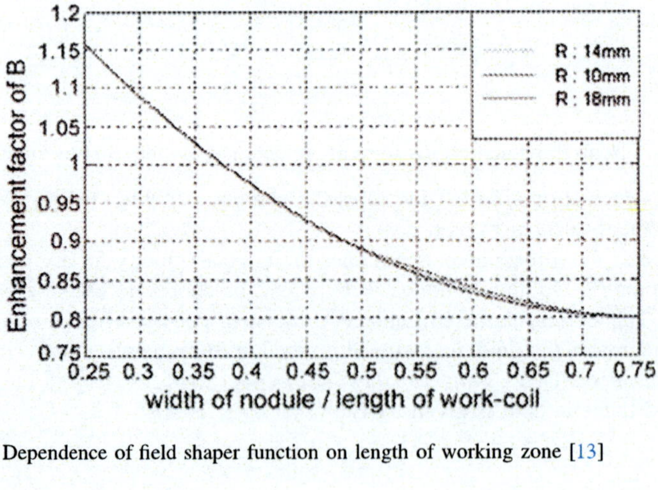

Fig. 7.17 Dependence of field shaper function on length of working zone [13]

Required current peak with its components can be described by Eqs. 7.7–7.10

$$I = i_{\mathrm{wz}} + 2i_{\mathrm{end}} \tag{7.7}$$

$$i_{\mathrm{wz}} = \phi.\frac{l}{2\pi\mu_o Rs}\left(1 - \frac{s}{2R}\right) + 2\Delta i \tag{7.8}$$

Δi—correction factor for the current is given by Eq. (7.9).

$$\Delta i = \frac{\phi}{\pi^2\mu_o R}(1 - \ln(2)) \tag{7.9}$$

$$i_{\mathrm{end}} = \frac{\phi}{\pi^2\mu_o R}\ln\left(\frac{\pi R}{2s}\right) - C_0 + D \tag{7.10}$$

A—surface between field shaper and workpiece, l—length of the work-zone = 15 mm, R—outer radius of flyer workpiece, s—standoff distance, C_0—Euler coefficeint = 0.57, D = 1, μ_0—permeability of vacuum.

Internal diameter of field shaper = 27 mm; insulation layer thickness = 1 mm.

To ensure correct alignment of the workpieces, the flyer workpiece will be chosen to match the outer diameter. With the known value of magnetic field, the required value of peak current can be determined.

Due to the magnetic field present in the region, magnetic pressure gets created. The change in the magnetic pressure causes radial forces to be exerted. Difference in the magnetic fields inside and outside of the workpiece causes the gradient in the magnetic pressure, and thus, a force which acts on the flyer tube.

This magnetic pressure is given by Eq. (7.11)

$$P(t) = \frac{1}{2\mu_0}\left[B_{\mathrm{gap}}(t)^2 - B_{\mathrm{diff}}(t)^2\right] \tag{7.11}$$

B_{gap}—magnetic field outside the workpiece, B_{diff}—magnetic field diffused through the workpiece.

If the magnetic field diffused through the workpiece is unknown, the pressure can also be found using the material properties based on the Lorentz force which acts on a current carrying body in presence of a magnetic field as in Eq. (7.12)

$$P = i_{\mathrm{d}} \cdot B_{\mathrm{gap}} \tag{7.12}$$

i_{d}—current density in workpiece, P—magnetic pressure (N/m^2).

Current density in the workpiece and the magnetic field are caused by the current in the solenoid and can be calculated using Maxwell's equation, and thus, a new expression for magnetic pressure Eq. (7.13):

$$P = \left(\frac{B_0^2}{2\pi}\right)\left(1 - e^{-\frac{2t}{\delta}}\right) \tag{7.13}$$

$B_0 = B_{gap}$ = magnetic field at outer surface of the tube, t—plate thickness, σ—electrical conductivity (m/Ω), μ—permeability (H/m), f—frequency of current (HZ).

Skin depth is the depth through which the interaction of both the magnetic fields is limited within the workpiece (flyer). Due to this skin effect, the repulsive magnetic field produces electromagnetic forces that exert the requisite pressure on the flyer. It should be smaller compared to the material thickness

- to maximize the magnetic pressure
- to maximize the influence of eddy currents, so that the current is concentrated at the interface and the magnetic field does not diffuse through the workpiece.

Magnetic field is observed to decline exponentially in the workpiece. Magnetic pressure is thus determined using the two components of magnetic field.

The required pressure can be considered as two components: one required for circumferential deformation of the tubular workpiece, given by Eq. (7.14)

$$P_1 = \frac{\alpha 2\sigma_y t}{R} \tag{7.14}$$

α—factor representing limits on deformation = 1, generally, $_y$—yield strength of the material (N/m^2), t—thickness of workpiece (m), R—radius of workpiece (m).

Second component of pressure is used to accelerate the working zone of the flyer tube to cause bonding at impact with the target tube. Assuming linear acceleration, this pressure is given by Eq. (7.15).

$$\text{Pressure } P_2 = m.\frac{a}{A} = t.\rho.\frac{v_c^2}{2s} \tag{7.15}$$

Standoff distance is given by Eq. (7.16)

$$s = \frac{v_c^2}{2a} \tag{7.16}$$

s—bandgap, v_c—impact velocity, a = acceleration of the flyer tube, m—mass per unit lenght of the outer tube, A—surface area of interaction of the flyer tube per unit of length (m^2/m), ρ—density, s—standoff distance.

Total pressure is given by Eq. (7.17).

$$P = P_1 + P_2 = \frac{\alpha * 2\sigma_y * t}{R} + \frac{t * \rho * v_c^2}{2s} \tag{7.17}$$

Required pressure leads to a value of the magnetic field given by Eq. (7.18).

$$P = \frac{B_o^2}{2\mu_0}\left(1 - e^{-\frac{2t}{\delta}}\right) \tag{7.18}$$

The magnetic field density gives the value of required peak current *I*.

Total time for which the workpiece is accelerated is given by Eq. (7.19).

$$t_{acc} = \frac{T}{2} - 2 * t1 \tag{7.19}$$

T—timeperiod, *t*1—time at which workpiece starts accelerating.

Assuming constant acceleration, time taken by the flyer piece to reach the target piece is given by Eq. (7.20):

$$t_{fly} = \sqrt{\left(\frac{2s}{a}\right)} \tag{7.20}$$

s—standoff distance, *a*—areas of acceleration.

Required voltage level is given by Eq. (7.21).

$$V = \frac{I_{peak}}{K}; \quad K \text{ given by } \frac{V_{max}}{I_{max}} \tag{7.21}$$

The influence of voltage level on the impact velocity, found using the above formulae, is represented by the set of curves in Fig. 7.18.

Based on the above equations, it can be inferred that the material properties have an influence on the primary current and also on the magnetic field.

7.7.2 Design Calculations II

- Pulse generator: capacitance = 160 μF, inductance = 1.5 μH, V_{max} = 20 kV, I_{max} = 500 kA, *f* =10 kHz [14]
- Material properties:
  - Yield strength—120 N/mm^2
  - Density—2700 kg/m^3
  - Conductivity—1.74 * 107 S/m
  - Skin depth—3.82 mm (using Eq. 7.1)
  - Thickness—2 mm
  - Length of working zone = 15 mm
  - Insulation layer thickness—1 mm

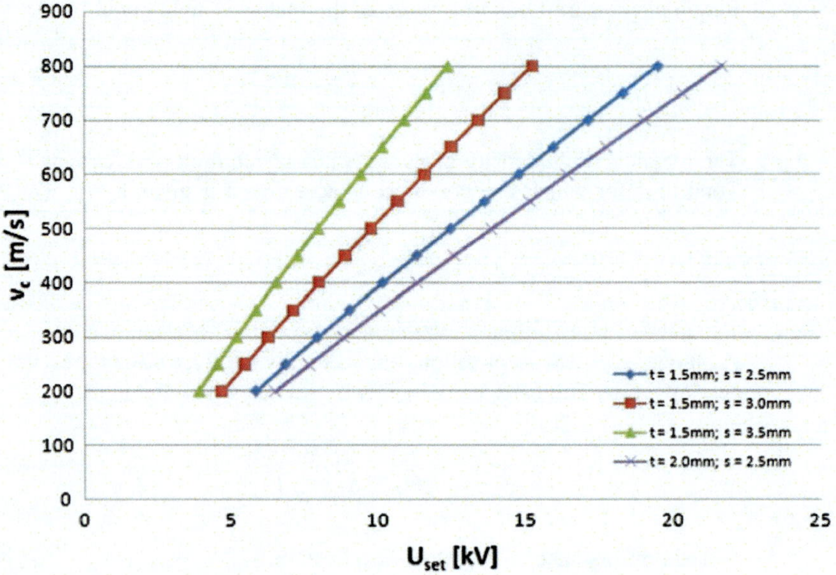

Fig. 7.18 Influence of applied voltage, plate thickness, and standoff distance on impact velocity [14]

– Radius of workpiece—12 mm
– Flyer workpiece diameter—25 mm.

Process parameters for Al flyer tube are found with the condition that the time taken in the acceleration is more than time of constant acceleration. This is to enable the flyer to reach the desired impact velocity with the required pressure.

- Impact velocity—200 m/s
- Standoff distance—2 mm
- For this impact velocity, the total pressure required can be found using Eq. (7.18). Total required pressure = 47 MPa
- Magnetic field density will thus be given by Eq. (7.6) = 13.48 T
- Flux: Magnetic flux density * area
- Area—Surface area between field shaper and workpiece is given by Eq. (7.22)

$$A = \pi\left(R_{\text{coil}}^2 - R^2\right) \tag{7.22}$$

- $A = 8.17 * 10^{-5}$ m^2 = 0.8 m^2
- Required current peak I = Current in the working zone + current in end zone = 79 kA
- Required voltage: $I_{peak} * V_{max}/I_{max} = 3.6\,kV$

7.7.3 *Design Case Study III*

Kumar et al. [1] worked on the approaches and engineering calculations required to effectively use the actuator in EMPW of flat components.

Figure 7.19 shows a schematic diagram of the discharge circuit. The circuit consists of a capacitor for a supply of electrical energy, a discharge gap switch, and an E-shaped one-turn flat coil. The two plates are placed above the coil with a small space between them.

Attempts were made to make a low-inductance discharge circuit that can generate a high-density magnetic flux around the coil area.

The Al work sheet: 25 mm * 25 mm * 1 mm
Stainless steelwork sheet: 35 mm * 35 mm * 2 mm
Distance between flyer plate and parent plate before start of welding: 1 mm.

The 0.1–0.3 mm-thick insulating sheets were loaded between the coil surface and the overlapped ends of the workpiece sheet.

The capacitor bank consists of 16 capacitors of 4 μF/75 kV in parallel. It is connected to the gap switch and one-turn coil by a low inductance transmission line

The flat E-shaped Cu coil thickness is 20 mm as shown in Fig. 7.20, and the inductance of the coil is 0.02 nH. Circuit is designed to keep the inductance value low to enable swift welding.

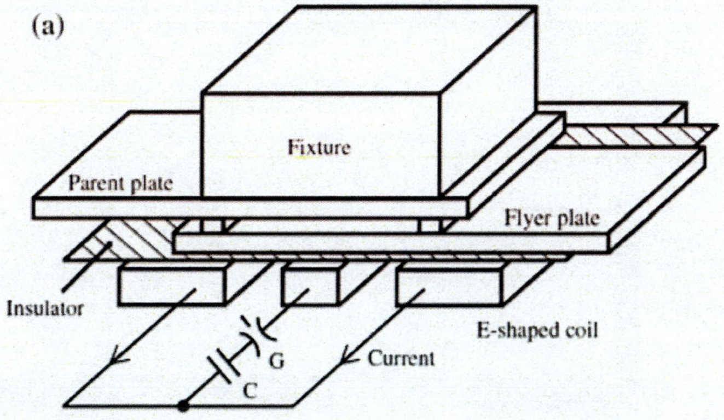

Fig. 7.19 Schematic of MPW discharge circuit [1]

Fig. 7.20 3D model of the coil [10]

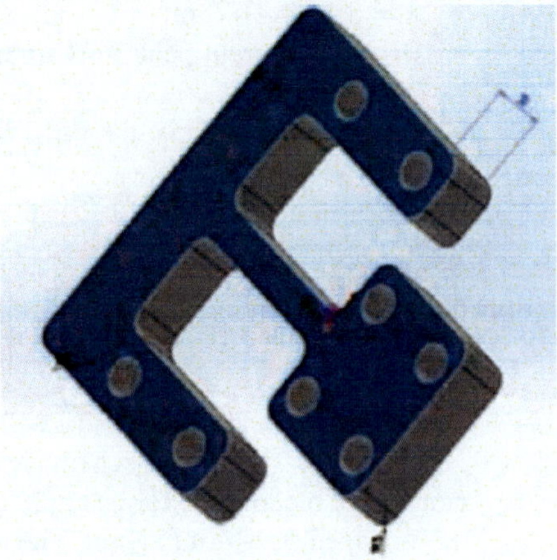

The web of 10 mm wide and 50 mm long helps increase the current density due to increase in the current concentration J. The web of the coil was supported with a specially designed and fabricated nested-type nylon fixture to avoid its bending in the process of welding of the sheets.

An increase in current density gives a corresponding increase in the resultant Lorentz force that generates stronger impact between the sheets to be welded.

The Lorentz force is given by: $F = J \times B$; where F is the Lorentz force in N, B the magnetic flux generated by the coil in Tesla (T), and J is in A/m^2.

Electromagnetic force generated to drive the flyer plate toward the parent plate is 14.36 kN.

Velocity of the flyer plate achieved is 92 m/s.

A typical current waveform is shown in Fig. 7.21.

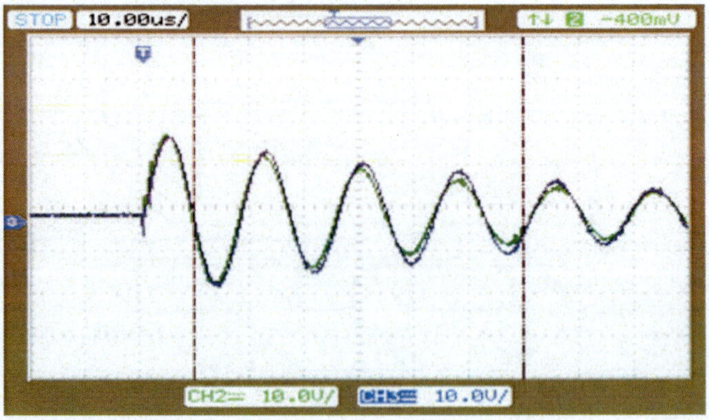

Fig. 7.21 Current waveform at 9 kJ discharge energy [1]

This current signal was obtained at 9 kJ discharge by using a magnetic probe. The current signal shows that a damping and oscillating current flows through a one-turn coil for the duration of about 50 μs. The maximum current was measured at about 280 kA.

7.7.4 Design Case Study IV

Miranda et al. [4] studied the following two cases of dissimilar welding and presented their analysis.

Case I For the joining of aluminum and titanium alloys in sheet form having thickness 0.5–1 mm, the maximum observed current was 150 kA, if a tank circuit of capacitor bank of 100 μF/10 kV and an inductance of 0.02 μH was used with an energy discharge of 1.2 kJ.

The interface between Al and Fe/Ti/Mg was observed to have a wave like formation as shown in Fig. 7.22, and no defects or intermetallic were observed at the interface.

For the welding of sheets having different thicknesses, like that of aluminum and steel having thickness of 1 mm and 0.25 mm, respectively, the capacitor bank was required to be charged with 10 kJ at 10 kV. And total circuit inductance introduced by a copper coil such that the total inductance of the circuit was 0.7 μH. To ensure that the flyer material's skin depth is less than its thickness, AC frequency was adjusted to 18.5 kHz.

Case II To study the effect of impact velocity on the dissimilar material bond, experimentation was done with the welding of pure Al(99.5) to TiAl6V4. Conclusions were made that the increase in the impact velocity produces a higher ratio of welded area and the contact surface, as shown in Fig. 7.23 and also that for lower impact velocity (10–25 m/s), no effective bond is formed.

The bonding interface in Fig. 7.24 shows the microfractures parallel to the contact surface of the weld in the micrograph analysis at the weld interfaces, for an impact velocity of 130 m/s. The condition at the joint deteriorates with an increase

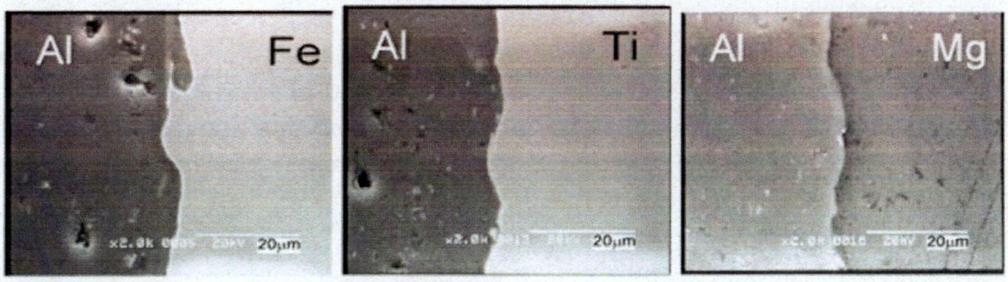

Fig. 7.22 Interface between Al and Fe/Ti/Mg [4]

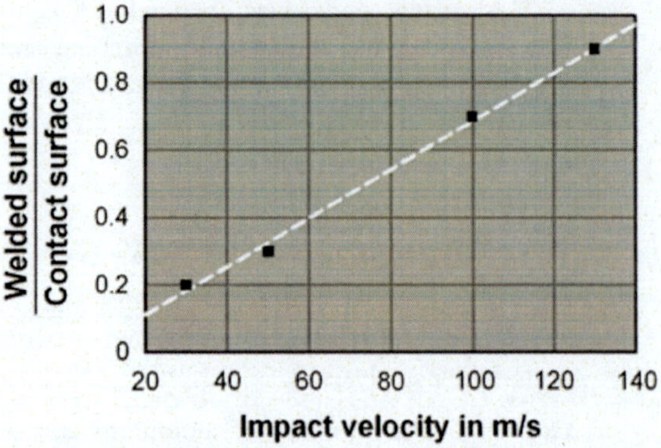

Fig. 7.23 Variation of welded surface-to-contact surface ratio [4]

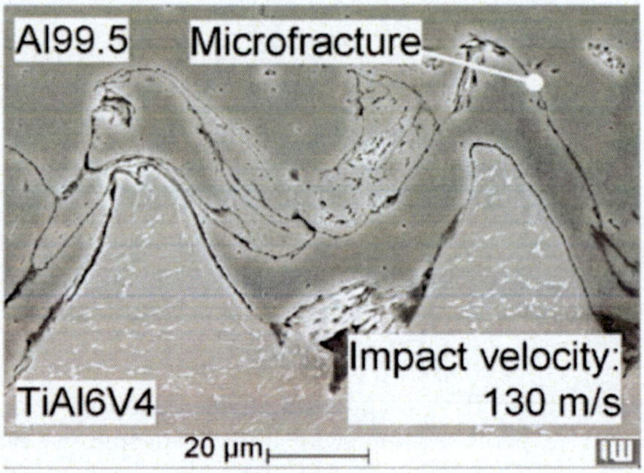

Fig. 7.24 Bonding interface [4]

in the impact velocity, thus indicating the optimum range of impact velocity that lies between 100 and 130 m/s.

The experiments were also conducted with the materials in the form of tubes having the following dimensions:

- Aluminum: diameter= 20 mm, thickness= 1 mm
- Titanium: diameter= 15 mm, thickness= 2.5 mm.

The workpieces were positioned coaxially with an initial gap of 1.5 mm in the compression coil. Based on the observations, 100 m/s was found to be the optimum impact velocity with an energy of 500 J. A wavy interface of 4–6 mm amplitude was found with the discharge energy of 1000 J, and it was also observed that a good weld results with an increased impact velocity.

7.7.5 Design Calculations—Weldability Curve

The damped sinusoidal discharge current and consequently time-dependent magnetic pressure poses difficulties in the accurate analytical modeling of MPW process. The magnetic pressure varies based upon the axial and circumferential position of a field shaper that may be used to increase the magnetic field intensity.

Also, complex deformation behavior and high-speed deformation of the workpieces adds to the problem of finding equations that have reasonable accuracy, as well as sufficient simplicity.

Analogy between MPW and explosion welding can be used to find the collision velocity depending on the materials. Grignon et al. [15] worked with various equations given by different researchers to arrive at the weldability curve for 6061 T0 Al alloy shown in Fig. 7.25.

The shaded area in the above plot represents the optimum set of values of welding velocity and collision angle for achieving an efficient weld. Required acceleration of the flyer plate can be then found by the optimum value of welding velocity for the standoff distance. This value of acceleration can then be used to find the required magnetic pressure p, using the general relation given in Eq. (7.23):

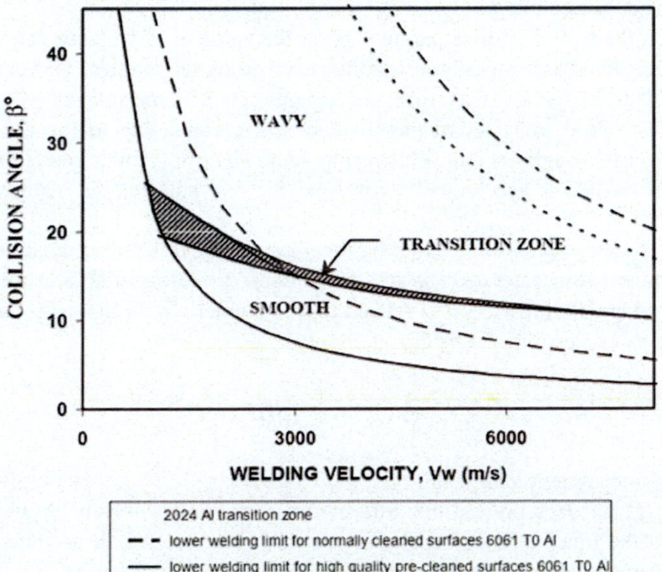

Fig. 7.25 Weldability window for 6061 T0 Al alloy [15]

$$\text{Acceleration } a = -\frac{1}{\rho} * \frac{\mathrm{d}P}{\mathrm{d}z}; \tag{7.23}$$

ρ—material density, dP—pressure gradient, dz—standoff distance.

An RLC circuit was used to model the weld circuit components. The discharge current from the capacitor bank is given as an exponentially damped sine wave described in Eq. (7.24).

$$i(t) = \frac{V_0}{\omega_c L} * e^{-\xi t} * \sin(\omega_c t) \tag{7.24}$$

V_0—capacitor bank voltage, ω_c—circuit frequency, ξ—damping factor.

Damping factor and circuit frequency are functions of R, L, and C and thus govern the functioning of the circuit.

Capacitance C is the total capacitance depending on the number of charged capacitors.

Inductance L shows a significant variation based on the radial distance between the field shaper and the flyer tube. This variation is generally neglected to simplify the analytical model.

On performing the current measurements for a certain set of geometries and materials for each field shaper, the values of frequency and damping factor can be determined. These values can be further used to calculate the values of resistance R and inductance L.

Influence of various parameters on the value of inductance can be analyzed and represented as a function using the curve fitting techniques. This can further help to predict the current waveform for various sets of experiments.

An axial transient magnetic field gets created due to the damped oscillating current through the coil. This results in an electromagnetic force acting on the flyer tube, which is accelerated away from the coil and collides rapidly with the inner tube.

Required pressure exerted by the magnetic field is the summation of the pressure required to accelerate (P_a) and the pressure to deform the flyer tube (P_d).

Magnetic pressure P is related to the magnetic field intensity outside and inside the workpiece, given by Eq. (7.25) [9]:

$$p = \frac{1}{2} * \mu\left(H_0^2 - H_i^2\right) : \tag{7.25}$$

μ—magnetic permeability.

H_0 and H_i represent the differing magnetic field intensity between the flyer tube and the inner tube in comparison with that between coil and the flyer tube. This difference is because of the shielding effect.

The magnetic pressure is also given by the expression (7.26):

$$p = \frac{1}{2\mu} B_0 \left(1 - e^{\frac{2T}{\delta}}\right) \text{with skin depth } \delta = \sqrt{\frac{2}{\omega\mu\kappa}} \tag{7.26}$$

Using the value of the pressure, strength of the magnetic field can be obtained.

The magnetic pressure which is obtained due to damped sinusoidal current will also be a function of time, and thus, the acceleration will also not be constant. This impedes the establishment an analytical expression for accelerations as a function of time.

7.7.6 MPW: Electrical Model

Thibaudeau [11] presented an analytical model of a uniform pressure actuator (UPA) to maximize the magnetic pressure and workpiece velocity. UPA is considered to be used as the work coil since it has appropriate pressure distribution over a larger area and is robust as compared to single turn coils. The magnetic pressure and the workpiece velocity are predicted to ensure the sufficient impact velocities for MPW. The analytical model which can be used to design the process parameters can be divided into stages based on the various interaction involved as shown in schematic Fig. 7.26.

For lightweight automotive applications, welding of flat sheets is required. Design of the working coil is a critical factor in the MPW process since the weld quality depends upon it. Larger turn coil results in a stronger magnetic field and also higher magnetic pressure. Having lesser number of turns allows for a shorter rise time to peak current.

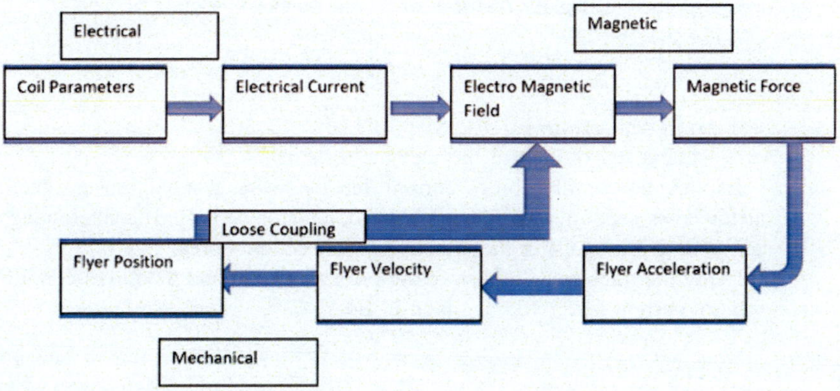

Fig. 7.26 Schematic of analytical process

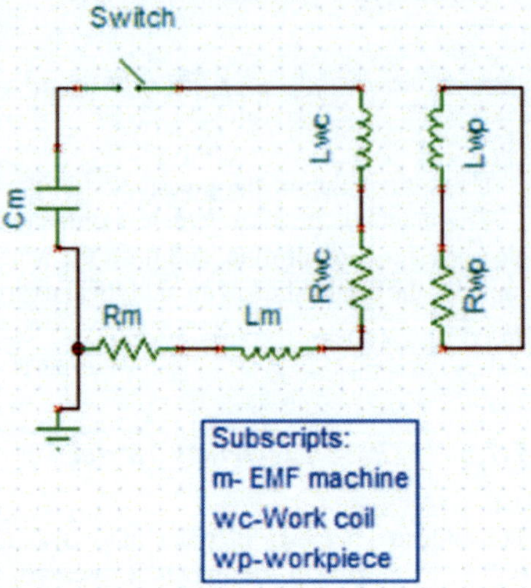

Fig. 7.27 Electrical model

Electrical model is represented as a combination of *R*, *L*, *C* elements [11] including the mutual inductance between the coil, and the workpiece as show in Fig. 7.27 is given by Eq. (7.27).

$$M = k\sqrt{L_C L_w} \tag{7.27}$$

For typical tubular and sheet workpieces, inductance of workpiece is negligible, thus mutual inductance can be neglected.

*R*m and *L*m can be determined experimentally by recording electrical response to a shorted load. *R*c and *L*c are calculated from the device geometries

Because of the oscillating current, skin effect produces an exponentially decaying current density distribution given by Eq. (7.28)

$$J = J_s e^{-\frac{d}{\delta}} \tag{7.28}$$

δ—skin depth from Eq. (7.26),

ω—frequency of current through the resistor.

Exponential distribution of current density causes a reduction in effective conductor cross section. Resistance of the conductor subjected to alternating current can be determined using the effective cross-sectional area.

Using the transient response of the primary circuit and the initial conditions, the primary current is derived as given in Eq. (7.29):

$$I(t) = V_0\sqrt{LC} * e^{-\omega\xi t} * \frac{\sin(\omega t)}{\sqrt{1-\zeta^2}}; \tag{7.29}$$

ξ—damping factor, ω = 2*πf*, *f*—ringing frequency, C—capactiance, L—inductance, V_0—input voltage.

Natural frequency of the circuit is given by Eq. (7.30).

$$\omega_n = 1/\sqrt{LC} \tag{7.30}$$

Damping ratio of the circuit is given as Eq. (7.31).

$$\zeta = \frac{R}{2} * \sqrt{\frac{C}{L}} \tag{7.31}$$

Assuming the value of capacitances from the standard values, further the values of *L* and *R* can be calculated.

Capacitor voltage can then be calculated by integrating the current out of the capacitor as in Eq. (7.32).

$$v_{c(t)} = \int(-i_p(t)\mathrm{d}t + v_{co} \tag{7.32}$$

Lorentz force is created due to the interaction of magnetic flux density and the current and is given in Eq. (7.33).

$$F = J * B; \quad J = -\frac{\partial H}{\partial y}; \quad B = \mu H - \rightarrow F = -\mu H\frac{\partial H}{\partial y} = -\frac{1}{2} * \mu\frac{\partial H^2}{\partial y} \tag{7.33}$$

The body force *F* can be integrated through the thickness of the workpiece to determine an effective pressure acting on the workpiece surface given by Eq. (7.34).

$$\text{Pressure } P = \int_{y1}^{y2} F\mathrm{d}y = \frac{1}{2} * \mu\left(H_{\text{gap}}^2 - H_{\text{pen}}^2\right) \tag{7.34}$$

H_{gap}—magnetic field strength in the gap region;

H_{pen}—magnetic field strength in the penetrated region.

If penetrated field strength is neglected due to skin effect, then the magnetic pressure is given by Eq. (7.35):

$$P_{\text{m}} = 0.5 * \mu k H_{\text{gap}}^2 \tag{7.35}$$

k = coefficient of magnetic coupling = 0.7 for Al and 0.55 for SS

The design parameters accounted for in this section for various geometrical profiles and materials, apart from the factors considered, also depend upon the geometry and placement of the workpieces inside the weld assembly as they influence the weld angle. Utilizing the arrived relations gives a broad range of the variable parameters and knowledge of the factors that may influence the required voltage level.

7.8 Conspectus of Design Studies in Magnetic Pulse Welding

This chapter deals with design aspects of MPW, an innovative weld solution showing great potential to replace some conventional processes in automobile industries. This process also presents itself as an alternative for dissimilar material joints. MPW is still in its nascent stage though it is being sporadically employed in a few industries. Having realized the futuristic scope and demand for this process, this chapter attempts to present design details for MPW. The chapter presents information on the process phenomena, applications, preferable materials, kinematics of the process, basic design procedures and calculation, and the various numerical methods adopted for process parametric optimizations. Case studies have been discussed to help readers relate their understanding to pragmatic technical scenarios involved during the employment of MPW for product fabrications.

Whatever has been dealt within this volume on welding design for advanced joining processes is only a tip of the iceberg. The exhaustive coverage of this multifarious continuum cannot be confined with this volume.

References

1. Kumar R, Sahoo S, Sarkar B, Shyam A (2017) Development of electromagnetic welding facility of flat plates for nuclear industry. In: J Phys Conf Ser 823(1):012039. https://doi.org/10.1088/1742-6596/823/1/012039. IOP Publishing
2. Zhang Y, Babu SS, Daehn GS (2010) Interfacial ultra- fine grained structures on aluminum alloy 6061 joint and copper alloy 110 joint fabricated by magnetic pulse welding. J Mater Sci 45(17):4645–4651. https://doi.org/10.1007/s10853-010-4676-0
3. Hahn M, Weddeling C, Lueg-Althoff J, Tekkaya AE (2016) Analytical approach for magnetic pulse welding of sheet connections. J Mater Process Technol 230:131–142. https://doi.org/10.1016/j.jmatprotec.2015.11.021
4. Miranda RM, Tomás B, Santos TG, Fernandes N (2014) Magnetic pulse welding on the cutting edge of industrial applications. Soldagem & Inspeção 19(1):69–81. https://doi.org/10.1590/S0104-92242014000100009
5. https://www.bmax.com/magnetic-pulse-welding/
6. Psyk V, Scheffler C, Linnemann M, Landgrebe D (2017) Process analysis for magnetic pulse welding of similar and dissimilar material sheet metal joints. Procedia Engineering 207:353–358. https://doi.org/10.1016/j.proeng.2017.10.787
7. Kapil A, Sharma A (2015) Magnetic pulse welding: an efficient and environmentally friendly multi-material joining technique. J Clean Prod 100:35–58. https://doi.org/10.1016/j.jclepro.2015.03.042

8. Khanolkar GR (2015) Simulation of magnetic pulse welding with varying air gap in tubular jobs using FEM (Doctoral dissertation)
9. Broeckhove J, Willemsens L, Faes K (2010) Magnetic pulse welding. Sustain Construct Design 1(1):21
10. Psyk V, Gershteyn G, Barlage B, Weddeling C, Albuja B, Brosius A, Tekkaya AE, Bach FW (2011) Process design for the manufacturing of magnetic pulse welded joints. In: Key engineering materials (Vol. 473). Trans Tech Publications, pp 243–250
11. Thibaudeau EH (2013) Development of analytical and experimental tools for magnetic pulse welding
12. Kamal M, Daehn GS (2007) A uniform pressure electromagnetic actuator for forming flat sheets. J Manuf Sci Eng 129(2):369–379. https://doi.org/10.1115/1.2515481
13. Bahmani MA, Niayesh K, Karimi A (2009) 3D Simulation of magnetic field distribution in electromagnetic forming systems with field-shaper. J Mater Process Technol 209(5):2295–2301
14. Loncke K (2009) An exploratory study into the feasibility of magnetic pulse welding. Department of Mechanical Construction and Production, Ghent University, Ghent, pp 1–150
15. Grignon F, Benson D, Vecchio KS, Meyers MA (2004) Explosive welding of aluminum to aluminum: analysis, computations and experiments. Int J Impact Eng 30(10):1333–1351